HERPETOLOGY OF AFRICA:
A Checklist and Bibliography of the Orders Amphisbaenia, Sauria and Serpentes

Herpetology of Africa:

A Checklist and Bibliography of the Orders Amphisbaenia, Sauria and Serpentes

by
KENNETH R. G. WELCH

ROBERT E. KRIEGER PUBLISHING COMPANY
MALABAR, FLORIDA
1982

Original Edition 1982

Printed and Published by
ROBERT E. KRIEGER PUBLISHING COMPANY, INC.
KRIEGER DRIVE
MALABAR, FLORIDA 32950

Library of Congress Cataloging in Publication Data

Welch, Kenneth R. G.
  Herpetology of Africa.

  Bibliography: p.
  Includes indexes.
    1. Reptiles—Africa.   2. Reptiles—Africa—Bibliography.
  I. Title.
  QL662.A1W44   597.9'096   81-17233
  ISBN 0-89874-428-8   AACR2

To
Maria A. and Sandor B. (Stockholm)
and the people of
Furuviksparken
Gavle
and
Tradgardsforeningen, Goteborg.

PREFACE and ACKNOWLEDGEMENTS

Since early childhood I have had more than a casual interest in the reptiles of Africa and thus, not without love, have taken it upon myself to prepare a list of the reptile fauna of the continent as a whole in an attempt to help standardize the names of the listed fauna. In doing so I undoubtedly will expose myself to great criticism but should it prompt further field and taxonomic studies then the study of herpetology in Africa can only benefit.

In preparing the list and its preliminary bibliography I have leaned heavily on many shoulders and would like to extend my thanks to all including: E. N. Arnold (B.M.N.H., London); C. Gans (U.S.A.); F. Gibbons (secretary of the South Western Herpetological Society); A. G. C. Grandison (B.M.N.H., London); D. E. Hahn (U.S.A.) for his knowledge on the Scolecophidians; K. A. Harding for his early encouragement; B. Lanza (Italy); C. J. McCarthy (B.M.N.H., London); G. Pasteur (France) for his knowledge of the genus *Chalcides;* and J. Richards and A. S. Wright for their help in the search for references. It must be said that all final decisions were taken by myself and thus I alone am responsible for any mistakes.

K. R. G. Welch
15 February 1981

CONTENTS

Chapter 1

CLASSIFICATION

Introduction. Over the years various regional herpetofaunal lists have appeared for parts of Africa but no attempt has been made to list the fauna of the continent as a whole; herein including mainland Africa, Canary Islands, Madeira, Cape Verde Islands, the islands of the Gulf of Guinea, Socotra, Madagascar, the Mascarenes, Seychelles, Aldabra, Europa Island and Juan de Nova Island. Such a list is readily subject to much comment and criticism which would change the contents and thus included under each taxa are those references which will aid the reader to prepare synonymies, description, distribution and information for further studies.

Classification. The following is meant only as a note on the classification used in the main text. The Order Amphisbaenia follows Gans (1967). The Order Sauria follows that of Underwood (1971) after the removal of the Amphisbaenids and the inclusion of the Gerrhosaurids as a subfamily of Cordylidae. The classification of the Order Serpentes is here presented in a novel way (see Smith, Smith and Sawin 1977).

The order is divided into two suborders corresponding to the infraorders of McDowell (1974) with one, Alethinophidia, being further divided into two infraorders Henophidia and Caenophidia (see Underwood 1967). The arrangement of the infraorder Henophidia follows McDowell (1975, 1979) with the exception that herein I use the tribal name Calabarini for the genus *Calabaria* (see Underwood 1976) and Pythonini for the remaining genera of the subfamily Pythoninae. The infraorder Caenophidia is divided into three superfamilies though undoubtedly in the not too distant future an Underwood-ian classification will change the arrangement within this infraorder: Colubroidea for the family Colubridae; Elapsoidea for the families Elapidae, Oxyuranidae and Hydrophiidae (see Burger and Natsuno, 1974); and Viperoidea for the families Viperidae and Crotalidae.

1

In the following list the reader should note that only those taxa found in our area are included.

Order Amphisbaenia
    Family Trogonophidae
    Family Amphisbaenidae

Order Sauria
  Suborder Ascalabota
    Infraorder Gekkota
      Superfamily Gekkonoidea
        Family Gekkonidae
          Subfamily Diplodactylinae
          Subfamily Gekkoninae
    Infraorder Iguania
      Family Iguanidae
      Family Agamidae
      Family Chamaeleonidae
  Suborder Autarchoglossa
    Infraorder Scincomorpha
      Superfamily Scincoidea
        Family Scincidae
          Subfamily Feylininae
          Subfamily Acontinae
          Subfamily Scincinae
          Subfamily Lygosominae
      Superfamily Lacertoidea
        Family Lacertidae
        Family Cordylidae
          Subfamily Gerrhosaurinae
          Subfamily Cordylinae
    Infraorder Anguimorpha
      Superfamily Anguioidea
        Family Anguidae
          Subfamily Anguinae
      Superfamily Varanoidea
        Family Varanidae

Order Serpentes
  Suborder Scolecophidia
    Family Leptotyphlopidae
    Family Typhlopidae
  Suborder Alethinophidia
  Infraorder Henophidia
    Superfamily Booidea
      Family Boidae
        Subfamily Pythoninae
          Tribe Calabarini
          Tribe Pythonini
        Subfamily Boinae
          Tribe Boini
          Tribe Erycini
    Superfamily Bolyerioidea
      Family Bolyeriidae
  Infraorder Caenophidia
    Superfamily Colubroidea
      Family Colubridae
        Subfamily Aparallactinae
        Subfamily Atractaspidinae
        Subfamily Boiginae
        Subfamily Boodontinae
        Subfamily Colubrinae
        Subfamily Dasypeltinae
        Subfamily Dispholidinae
        Subfamily Geodipsadinae
        Subfamily Lycodontinae
        Subfamily Lycophidinae
        Subfamily Natricinae
        Subfamily Philothamninae
        Subfamily Psammophinae
        Subfamily Pseudaspidinae
        Subfamily 'Uncertain'
    Superfamily Elapsoidea
      Family Elapidae
        Subfamily Bungarinae
          Tribe Dendroaspini
          Tribe Najini
      Family Hydrophiidae
        Subfamily Hydrophiinae
    Superfamily Viperoidea
      Family Viperidae
        Subfamily Viperinae
        Subfamily Causinae

Chapter 2

FAMILY TROGONOPHIDAE

Order: Amphisbaenia
Family: Trogonophidae

Genus: AGAMODON Peters 1882
Species typica: *anguliceps* Peters
Reference: Gans and Pandit (1965)

Agamodon anguliceps anguliceps Peters 1882
    Distribution: South-central coast of Somalia
    Reference: Bonin (1965); Gans (1960c)

Agamodon anguliceps immaculatus Calabresi 1927
    Distribution: Somalia

Agamodon compressus Mocquard 1888
    Distribution: Central coastal Somalia
    Reference: Gans (1960c)

Genus: PACHYCALAMUS Gunther 1881
Species typica: *brevis* Gunther

Pachycalamus brevis Gunther 1881
    Distribution: Socotra
    Reference: Loveridge 1941c

Genus: TROGONOPHIS Kaup 1830
Species typica: *wiegmanni* Kaup

Trogonophis wiegmanni wiegmanni Kaup 1830
    Distribution: Eastern Morocco, northern Algeria and western Tunisia
    Reference: Gans (1960c); Loveridge (1941c); Pasteur and Bons (1960)

Trogonophis wiegmanni elegans (Gervais 1835)   *Amphisbaena*
    Distribution: Western Morocco
    Reference: Gans and Pasteur (1962)

Chapter 3

FAMILY AMPHISBAENIDAE

Order: Amphisbaenia
Family: Amphisbaenidae

Genus: ANCYLOCRANIUM Parker 1942
Species typica: *somalicus* Scortecci
Reference: Gans and Kochva (1966)

Ancylocranium barkeri barkeri Loveridge 1946
    Distribution: Tanzania
    Reference: Loveridge (1955), (1962)

Ancylocranium barkeri newalae Loveridge 1962
    Distribution: Tanzania

Ancylocranium ionidesi ionidesi Loveridge 1955
    Distribution: Tanzania
    Reference: Loveridge (1962)

Ancylocranium ionidesi haasi Gans and Kochva 1966
    Distribution: Rhondo Plateau, Tanzania

Ancylocranium somalicum somalicum (Scortecci 1931)   *Anops*
    Distribution: Somalia
    Reference: Loveridge (1941c)

Ancylocranium somalicum parkeri Gans and Kochva 1966
    Distribution: Somalia

Genus: BAIKIA Gray 1865
Species typica: *africana* Gray

Baikia africana Gray 1865
    Distribution: Northern Nigeria
    Reference: Dunger (1965), (1968b); Loveridge (1941c); Parker (1942)

Genus: BLANUS Wagler 1830
Species typica: *cinerea* Vandelli

Blanus cinereus cinereus (Vandelli 1797)   *Amphisbaena*
    Distribution: vicinity of Tangier, Morocco
    Reference: Arnold and Burton (1978); Bons (1963); Gans (1962)

Blanus cinereus mettetali Bons 1963
    Distribution: Morocco

Genus: CHIRINDIA Boulenger 1907
Species typica: *swynnertoni* Boulenger
Reference: Broadley and Gans (1978a); Gans and Rhodes (1967); Loveridge
        (1941c)

Chirindia ewerbecki ewerbecki Werner 1910
    Distribution: coastal Southern Province, Tanzania
    Reference: Loveridge (1962)

Chirindia ewerbecki nanguruwensis (Loveridge 1962)   *Amphisbaena*
    Distribution: Southern Province, Tanzania

Chirindia langi Fitzsimons 1939
    Distribution: Transvaal
    Reference: Fitzsimons (1943)

Chirindia mpwapwaensis (Loveridge 1932)   *Amphisbaena*
    Distribution: Tanzania

Chirindia orientalis (Sternfeld 1911)   *Amphisbaenula*
    Distribution: Tanzania

Chirindia rondoensis (Loveridge 1941)   *Amphisbaena*
    Distribution: Rondo and Makonde Plateaus, Tanzania
    Reference: Loveridge (1962)

Chirindia swynnertoni Boulenger 1907
    Distribution: Zimbabwe and southern Mozambique
    Reference: Broadley (1963b), (1964a); Fitzsimons (1943)
    Amphisbaena bushbyi (Cott): Fitzsimons (1943)
    Chirindia bushbyi Cott: Broadley (1963b); Loveridge (1941c)

Genus: CYNISCA Gray 1844
Species typica: *leucura* Dumeril and Bibron
Reference: Laurent (1947b); Loveridge (1941c)

Cynisca bifrontalis (Boulenger 1906)   *Amphisbaena*
    Distribution: Congo

Cynisca degrysi (Loveridge 1941)   *Placogaster*
    Distribution: Sierra Leone

Cynisca feae (Boulenger 1906)   *Placogaster*
    Distribution: Coastal Guinea Bissau
    Reference: Manacas (1955)

Cynisca gansi Dunger 1968
    Distribution: Nigeria

Cynisca haughi (Mocquard 1904)   *Amphisbaena*
    Distribution: Gabon
    Reference: Dunger (1968b)

Cynisca kigomensis Dunger 1968
    Distribution: Nigeria

Cynisca kraussi (Peters 1878)   *Amphisbaena*
    Distribution: Banda Hills, northwest Ashanti, Ghana
    Reference: Loveridge (1948)

Cynisca lamottei (Angel 1943)   *Amphisbaena*
    Distribution: Guinea
    Reference: Angel and others (1954)

Cynisca leonina (Muller 1885)   *Amphisbaena*
    Distribution: Guinea Bissau and Los Archipelago

Cynisca leucura (Dumeril and Bibron 1839)   *Amphisbaena*
    Distribution: Ivory Coast, Ghana, Togo, Benin and Nigeria
    Reference: Dunger (1968b); Loveridge (1952)

Cynisca liberiensis (Boulenger 1878)   *Ophioproctes*
    Distribution: Liberia

Cynisca mulleri (Strauch 1881)   *Amphisbaena*
    Distribution: Sierra Leone, Liberia, Ivory Coast and Ghana

Cynisca nigeriensis Dunger 1968
    Distribution: Northern Nigeria

Cynisca oligopholis (Boulenger 1906)   *Amphisbaena*
    Distribution: Guinea Bissau

Cynisca rouxae Hahn 1979
    Distribution: Ivory Coast

Cynisca schaeferi (Sternfeld 1912)   *Chirindia*
    Distribution: Cameroon

Genus: DALOPHIA Gray 1865
Species typica: *welwitschii* Gray
Reference: Broadley, Gans and Visser (1976); Laurent (1947b);
           Loveridge (1941c)

Dalophia angolensis Gans 1976
    Distribution: Angola and Zambia

Dalophia ellenbergeri (Angel 1920)   *Monopeltis*
    Distribution: Zambia

Dalophia gigantea (Peracca 1903)   *Monopeltis*
    Distribution: Zaire
    Reference: Witte (1954)

Dalophia longicauda (Werner 1915)   *Monopeltis*
    Distribution: Namibia, Botswana and Zimbabwe
    Reference:
    Monopeltis granti colobura Boulenger: Fitzsimons (1943, part)

Dalophia luluae (Witte and Laurent 1942)   *Monopeltis*
    Distribution: Zaire
    Reference: Laurent (1964b)

Dalophia pistillum (Boettger 1895)   *Monopeltis*
    Distribution: Namibia, Cape Province, Transvaal, Botswana, Zimbabwe,
            and Mozambique
    Reference: Broadley (1962a); Laurent (1964b)
    Monopeltis granti Boulenger: Fitzsimons (1943); Laurent (1964b)
    Monopeltis granti colobura Boulenger: Fitzsimons (1943, part)
    Monopeltis granti transvaalensis Fitzsimons: Fitzsimons (1943)
    Monopeltis jallae Peracca: Loveridge (1941c)

Dalophia welwitschii Gray 1865
    Distribution: Angola

Genus: GEOCALAMUS Gunther 1880
Species typica: *modestus* Gunther
Reference: Loveridge (1941c)

Geocalamus acutus Sternfeld 1912
    Distribution: Southeastern Kenya and northeastern Tanzania

Geocalamus modestus Gunther 1880
    Reference: Central Tanzania

Genus: LOVERIDGEA Vanzolini 1951
Species typica: *phylofiniens* Tornier

Loveridgea ionidesi (Battersby 1950)   *Amphisbaena*
    Distribution: Southern Tanzania
    Reference: Loveridge (1951a), (1955)

Loveridgea phylofiniens (Tornier 1899)   *Amphisbaena*
    Distribution: Tanzania
    Reference: Loveridge (1941c)

Genus: MONOPELTIS Smith 1848
Species typica: *capensis* Smith
Reference: Broadley, Gans and Visser (1976); Fitzsimons (1943); Gans and
            Lehman (1973); Loveridge (1941c); Witte (1954)

Monopeltis adercae Witte 1953
    Distribution: Zaire

Monopeltis anchietae (Bocage 1873)   *Lepidosternon*
    Distribution: Angola, Namibia and Botswana

Monopeltis capensis capensis Smith 1848
    Distribution: Cape Province, Orange Free State, Transvaal, Botswana,
              Namibia, Angola, southern Zimbabwe and southern
              Mozambique
    Reference: Broadley (1962a); Kritzinger (1946); Visser (1967b)
    Monopeltis decosteri Boulenger: Broadley (1963a); Fitzsimons (1943);
              Loveridge (1941c)

Monopeltis capensis rhodesianus Broadley, Gans and Visser 1976
    Distribution: Zimbabwe and adjacent Zambia, Malawi and Mozambique

Monopeltis galeata (Hallowell 1852)   *Phractogonus*
    Distribution: Gabon

Monopeltis guentheri Boulenger 1885
    Distribution: Congo and Zaire
    Reference: Gans and Latifi (1971)

Monopeltis jugularis Peters 1880
    Distribution: Cameroon, Rio Muni and Gabon

Monopeltis kabindae Witte and Laurent 1942
    Distribution: Zaire

Monopeltis leonhardi Werner 1910
    Distribution: Namibia, Botswana, Cape Province and Zimbabwe

Monopeltis luandae Gans 1976
    Distribution: Angola

Monopeltis perplexus Gans 1976
    Distribution: Angola

Monopeltis remaclei Witte 1933
    Distribution: Southern Zaire

Monopeltis scalper (Gunther 1876)   *Phractogonus*
  Distribution: Zaire
  Reference: Laurent (1954)
  Monopeltis scalper bulsi Laurent: Laurent (1956b)
  Monopeltis scalper gerardi Boulenger: Laurent (1964a); Witte (1954)

Monopeltis schoutedeni Witte 1933
  Distribution: vicinity of middle Zaire River, Congo and Zaire

Monopeltis sphenorhynchus sphenorhynchus Peters 1879
  Distribution: Transvaal, Natal and Mozambique
  Reference: Broadley (1962a), (1963a)
  Monopeltis capensis gazei Fitzsimons: Fitzsimons (1943); Loveridge
            (1941c)
  Monopeltis habenichti Fitzsimons: Fitzsimons (1943); Loveridge (1941c);
          · Manacas (1957)

Monopeltis sphenorhynchus mauricei Parker 1935
  Distribution: Namibia, Cape Province, Botswana, Zimbabwe and Zambia
  Reference:
  Monopeltis ocularis Fitzsimons: Broadley (1962a); Fitzsimons (1943)

Monopeltis vanderysti Witte 1922
  Distribution: Zaire and Angola
  Reference:
  Monopeltis vanderysti closei Laurent: Laurent (1954)
  Monopeltis vanderysti lujae Witte: Loveridge (1941c); Witte (1954)
  Monopeltis vanderysti vilhenai Laurent: Laurent (1954), (1964b)

Monopeltis zambezensis Gans and Broadley 1974
  Distribution: Zimbabwe

Genus: ZYGASPIS Cope 1885
Species typica: *quadrifrons* Peters
Reference: Broadley and Gans (1969); Fitzsimons (1943); Loveridge (1941c);
          Witte (1954)

Zygaspis niger Broadley and Gans 1969
  Distribution: Zambia and Angola

Zygaspis quadrifrons quadrifrons (Peters 1862)   *Amphisbaena*
  Distribution: Angola, northern Namibia east into Botswana

Zygaspis quadrifrons capensis (Thominot 1887)   *Amphisbaena*
    Distribution: Cape Province east to Mozambique and north to Zaire
    Reference: Broadley (1962a); Laurent (1964b)

Zygaspis quadrifrons dolichomenta (Witte and Laurent 1942)
                                                                   Amphisbaena
    Distribution: Zaire

Zygaspis violacea (Peters 1854)   *Amphisbaena*
    Distribution: Zimbabwe, Mozambique, Zululand and Transvaal
    Reference: Manacas (1957)
    Amphisbaena violacea vandami Fitzsimons: Fitzsimons (1943);
                                    Loveridge (1951a)

Chapter 4

FAMILY GEKKONIDAE

Order: Sauria
Suborder: Ascalabota
Infraorder: Gekkota
Superfamily: Gekkonoidea
Family: Gekkonidae
Subfamily: Diplodactylinae

Genus: AFROEDURA Loveridge 1944
Species Typica: *bogerti* Loveridge
Reference: Fitzsimons (1943); Loveridge (1947)

Afroedura africana africana (Boulenger 1888)   *Oedura*
    Distribution: Namibia
    Reference: Haacke (1965)

Afroedura africana tirasensis Haacke (1965)
    Distribution: Namibia

Afroedura karroica karroica (Hewitt 1925)   *Oedura*
    Distribution: South-central Cape Province
    Reference: Webb (1951)
    Oedura karroica wilmoti Hewitt: Fitzsimons (1943)

Afroedura karroica bogerti Loveridge 1944
    Distribution: Eastern central Angola

Afroedura karroica halli (Hewitt 1935)   *Oedura*
    Distribution: Northeastern Cape Province and Botswana

Afroedura namaquensis (Fitzsimons 1938)   *Oedura*
    Distribution: Cape Province

Afroedura nivaria (Boulenger 1894)   *Oedura*
    Distribution: Natal and eastern Cape Province
    Reference:
    Oedura amatolica Hewitt: Fitzsimons (1943)

Afroedura pondolia pondolia (Hewitt 1925)   *Oedura*
    Distribution: Coastal eastern Republic of South Africa
    Reference:
    Oedura langi Fitzsimons: Fitzsimons (1943)

Afroedura pondolia marleyi (Fitzsimons 1930)   *Oedura*
    Distribution: Zululand

Afroedura pondolia multiporis (Hewitt 1925)   *Oedura*
    Distribution: Transvaal

Afroedura tembulica (Hewitt 1926)   *Oedura*
    Distribution: Cape Province

Afroedura transvaalica transvaalica (Hewitt 1925)   *Oedura*
    Distribution: Transvaal

Afroedura transvaalica platyceps (Hewitt 1925)   *Oedura*
    Distribution: Zimbabwe

Genus: CHONDRODACTYLUS Peters 1870
Species typica: *angulifer* Peters

Chondrodactylus angulifer Peters 1870
    Distribution: Western Cape Province and Namibia
    Reference: Fitzsimons (1943); Giess (1965); Haacke (1976c); Loveridge
            (1947)

Genus: COLOPUS Peters 1869
Species typica: *wahlbergii* Peters

Colopus wahlbergii Peters 1869
    Distribution: Botswana and Namibia
    Reference: Fitzsimons (1943); Haacke (1976b); Loveridge (1947);
        Mitchell and Steyn (1967)

Genus: KAOKOGECKO Steyn and Haacke 1966
Species typica: *vanzyli* Steyn and Haacke

Kaokogecko vanzyli Steyn and Haacke 1966
    Distribution: Coastal northern Namibia

Genus: PALMATOGECKO Andersson 1908
Species typica: *rangei* Andersson

Palmatogecko rangei Andersson 1908
    Distribution: Coastal Namibia
    Reference: Brain (1958) (1962c); Fitzsimons (1943); Lawrence (1959);
        Loveridge (1947); Procter (1929); Webb (1951); Werner
        (1977)

Genus: PAROEDURA Gunther 1879
Species typica: *sanctijohannis* Gunther
Reference: Angel (1942); Dixon and Kroll (1974)

Paroedura androyensis (Grandidier 1867)   *Phyllodactylus*
    Distribution: Southeastern Madagascar

Paroedura bastardi (Mocquard 1900)   *Phyllodactylus*
    Distribution: Southern Madagascar

Paroedura gracilis (Boulenger 1896)   *Diplodactylus*
    Distribution: Eastern Madagascar

Paroedura guibeae Dixon and Kroll 1974
    Distribution: Southern Madagascar

Paroedura homalorhinus (Angel 1936)   *Phyllodactylus*
    Distribution: Northern Madagascar

Paroedura oviceps (Boettger 1881)   *Phyllodactylus*
    Distribution: Northern Madagascar

Paroedura pictus (Peters 1854)   *Diplodactylus*
    Distribution: Southern Madagascar

Paroedura sanctijohannis Gunther 1879
    Distribution: Comoro Islands

Paroedura stumpffi (Boettger 1878)   *Phyllodactylus*
    Distribution: Northern Madagascar and the Comoro Islands

Genus: PHELSUMA Gray 1825
Species typica: *cepedianus* Merrem
Reference: Loveridge (1942a) (1947) (1957)

Phelsuma abbotti abbotti Stejneger 1894
    Distribution: Aldabra

Phelsuma abbotti longinsulae Rendahl 1935
    Distribution: Seychelles
    Reference: Tansley (1961)

Phelsuma abbotti pulchra Rendahl 1939
    Distribution: Seychelles

Phelsuma astriata astriata Tornier 1901
    Distribution: Seychelles
    Reference: Crawford and Thrope (1979); Thorpe and Crawford (1979)

Phelsuma astriata astovei Fitzsimons 1948
    Distribution: Astove Island near Aldabra

Phelsuma barbouri Loveridge 1942
    Distribution: Madagascar

Phelsuma bimaculata Kaudern 1922
    Distribution: Madagascar
    Reference: Angel (1942)

Phelsuma cepediana (Merrem 1820)   *Gecko*
    Distribution: Mascarenes

Phelsuma dubia dubia (Boettger 1880)   *Pachydactylus*
    Distribution: Coastal Tanzania and Madagascar
    Reference: Angel (1942)

Phelsuma dubia comorensis Boettger 1913
    Distribution: Comoro Islands

Phelsuma flavigularis Mertens 1962
    Distribution: Madagascar

Phelsuma guentheri Boulenger 1885
    Distribution: Mascarenes

Phelsuma guttata Kaudern 1922
    Disribution: Madagascar

Phelsuma laticauda laticauda (Boettger 1880)   *Pachydactylus*
    Distribution: Madagascar, Comoroes and Seychelles
    Reference: Angel (1942)

Phelsuma laticauda angularis Mertens 1964
    Distribution: Madagascar

Phelsuma lineata lineata Gray 1842
    Distribution: Madagascar
    Reference: Angel (1942)

Phelsuma lineata bombetokensis Mertens 1964
    Distribution: Madagascar

Phelsuma lineata chloroscelis Mertens 1962
    Distribution: Madagascar

Phelsuma lineata dorsovitata Mertens 1964
    Distribution: Madagascar

Phelsuma lineata pusilla Mertens 1964
    Distribution: Madagascar

Phelsuma madagascariensis madagascariensis Gray 1831
    Distribution: Madagascar
    Reference: Crawford and Thorpe (1979); Demeter (1976); Gabe and
            Saint Girons (1965); Mertens (1953); Switak (1966); Thorpe
            and Crawford (1979)

Phelsuma madagascariensis kochi Mertens 1954
    Distribution: Madagascar

Phelsuma madagascariensis sundbergi Rendahl 1939
    Distribution: Seychelles

Phelsuma mutabilis (Grandidier 1869)   *Platydactylus*
    Distribution: Madagascar
    Reference: Angel (1942)
    Phelsuma micropholis Boettger: Angel (1942)

Phelsuma newtoni Boulenger 1884
    Distribution: Mascarenes

Phelsuma ocellata (Boulenger 1885)   *Rhotropus*
    Distribution: Northwestern Cape Province
    Reference: Fitzsimons (1943); Russell (1977)

Phelsuma parkeri Loveridge 1941
    Distribution: Pemba Island

Phelsuma quadriocellata (Peters 1883)   *Pachydactylus*
    Distribution: Madagascar

Phelsuma serraticauda Mertens 1963
    Distribution: Madagascar

Phelsuma standingi Methuen and Hewitt 1913
    Distribution: Madagascar

Phelsuma trilineata Gray 1842
    Distribution: Madagascar

Phelsuma vinsoni Mertens 1963
    Distribution: Mascarenes

Phelsuma v-nigra Boettger 1913
    Distribution: Comoro Islands
    Reference: Angel (1942)

Genus: PTENOPUS Gray 1865
Species typica: *garrulus* Smith
Reference: Brain (1962c); Haacke (1964)

Ptenopus carpi Brain 1962
    Distribution: Namibia

Ptenopus garrulus (Smith 1849)   *Stenodactylus*
    Distribution: Botswana and northwestern Transvaal west through south-
                ern Kalahari to Namibia and northwestern Cape Province
    Reference: Fitzsimons (1943); Giess (1965); Loveridge (1947)
    Ptenopus garrulus maculatus Gray: Fitzsimons (1943); Loveridge
                                (1947); Werner (1977)

Ptenopus kochi Haacke 1964
    Distribution: Namibia

Genus: QUEDENFELDTIA Boettger 1883
Species typica: *trachyblepharus* Boettger

Quendenfeldtia trachyblepharus (Boettger 1874)   *Gymnodactylus*
    Distribution: Morocco
    Reference: Gabe and Saint Girons (1972); Loveridge (1947); Pasteur and
                Bons (1960); Pellegrin (1925b) (1926a) (1926e) (1927a)

Genus: RHOPTROPUS Peters 1869
Species typica: *afer* Peters
Reference: Fitzsimons (1943); Loveridge (1947)

Rhoptropus afer Peters 1869
    Distribution: Namibia and southern Angola
    Reference: Brain (1962c); Werner (1977)

Rhoptropus barnardi Hewitt 1926
    Distribution: Northern Namibia and southern Angola
    Reference: Laurent (1964b)

Rhoptropus boultoni boultoni Schmidt 1933
    Distribution: Namibia and Mossamedes Province, Angola

Rhoptropus boultoni benguellensis Mertens 1938
    Distribution: Benguela Province, Angola
    Reference: Laurent (1964b)

Rhoptropus boultoni montanus Laurent 1964
    Distribution: Huila, Angola

Rhoptropus braconnieri (Thominot 1878)   *Dactychilikion*
    Distribution: Botswana

Rhoptropus bradfieldi bradfieldi Hewitt 1935
Distribution: Namibia
Reference: Werner (1977)

Rhoptropus bradfieldi diporus Haacke 1965
Distribution: Northern Namibia

Rhoptropus taeniostictus Laurent 1964
Distribution: Mossamedes Province, Angola

Genus: SAURODACTYLUS Fitzinger 1843
Species typica: *mauritanicus* Dumeril and Bibron
Reference: Loveridge (1947); Pasteur and Bons (1960)

Saurodactylus fasciatus Werner 1931
Distribution: North and western Morocco
Reference: Bons and Pasteur (1957a); Hediger (1935) (1937)

Saurodactylus mauritanicus mauritanicus (Dumeril and Bibron 1836)
Gymnodactylus
Distribution: Algeria and northeastern Morocco
Reference: Pellegrin (1926a) (1927a); Preiss (1977)

Saurodactylus mauritanicus brosseti Bons and Pasteur 1957
Distribution: Morocco

Subfamily: Gekkoninae

Genus: AILURONYX Fitzinger 1843
Species typica: *seychellensis* Dumeril and Bibron

Ailuronyx seychellensis (Dumeril and Bibron 1836)   *Phyllodactylus*
Distribution: Seychelles

Ailuronyx trachygaster (Dumeril 1851)   *Platydactylus*
Distribution: Madagascar
Reference: Angel (1942)

Genus: ANCYLODACTYLUS Muller 1907
Species typica: *spinicollis* Muller

Ancylodactylus spinicollis Muller 1907
    Distribution: Cameroon
    Reference: Loveridge (1947)

Genus: BUNOPUS Blanford 1874
Species typica: *tuberculata*
Comment: The use of this genus follows that of Szczerback and Golubev
        (1977)

Bunopus blanfordi Strauch 1887
    Distribution: Egypt
    Reference: Loveridge (1947)

Genus: CNEMASPIS Strauch 1887
Species typica: *boulengeri* Strauch
Reference: Loveridge (1947)

Cnemaspis africana africana (Werner 1895)   *Gymnodactylus*
    Distribution: Central Kenya south to eastern Tanzania

Cnemaspis africana elgonensis Loveridge 1936
    Distribution: Western Kenya and Uganda

Cnemaspis africana kobleri Mertens 1937
    Distribution: Cameroon

Cnemaspis africana occidentalis Angel 1943
    Distribution: Guinea

Cnemaspis africana quattuorseriata (Sternfeld 1912)   *Gonatodes*
    Distribution: Kenya and Tanzania west into Zaire

Genus: CYRTODACTYLUS Gray 1827
Species typica: *pulchellus*
Comment: The use of this genus follows Szczerbak and Golubev (1977),
        with their subgenera being treated as full genera.

Cyrtodactylus scaber (Heyden 1827)   *Stenodactylus*
    Distribution: Egypt, Sudan and Ethiopia
    Reference: Loveridge (1947); Marx (1968)

Genus: EBENAVIA Boettger 1878
Species typica: *inunguis* Boettger

Ebenavia inunguis Boettger 1878
    Distribution: Comoro Islands, Madagascar and the Mascarenes
    Reference: Angel (1942)

Genus: GECKOLEPIS Grandidier 1867
Species typica: *typica* Grandidier
Reference: Angel (1942)

Geckolepis anomala Mocquard 1909
    Distribution: Madagascar

Geckolepis maculata Peters 1880
    Distribution: Madagascar and the Comoro Islands

Geckolepis petiti Angel 1942
    Distribution: Madagascar

Geckolepis polylepis Boettger 1893
    Distribution: Madagascar

Geckolepis typica typica Grandidier 1867
    Distribution: Madagascar

Geckolepis typica modesta Methuen and Hewitt 1913
    Distribution: Madagascar

Genus: GECKONIA Mocquard 1895
Species typica: *chazaliae* Mocquard

Geckonia chazaliae Mocquard 1895
    Distribution: Coastal Mauritania, Western Sahara and Morocco
    Reference: Loveridge (1947); Pasteur and Bons (1960)

Genus: HEMIDACTYLUS Oken 1817
Species typica: *mabouia* Moreau de Jonnes
Reference: Loveridge (1947)

Hemidactylus albopunctatus Loveridge 1947
    Distribution: Somalia and northern Kenya

Hemidactylus ansorgii Boulenger 1901
    Distribution: Nigeria west to Liberia

Hemidactylus aporus Boulenger 1906
    Distribution: Annobon Island, Gulf of Guinea

Hemidactylus arnoldi Lanza 1978
    Distribution: Northern Somalia

Hemidactylus barodanus Boulenger 1901
    Distribution: Ethiopia and Somalia
    Reference: Parker (1942)

Hemidactylus bavazzanoi Lanza 1978
    Distribution: Southern Somalia

Hemidactylus benguellensis Bocage 1893
    Distribution: Angola
    Reference: Broadley (1977d)

Hemidactylus bouvieri bouvieri (Bocourt 1870)   *Emydactylus*
    Distribution: Cape Verde Islands

Hemidactylus bouvieri boavistensis Boulenger 1906
    Distribution: Boa Vista and Ile Sal, Cape Verde Islands
    Reference:
    Hemidactylus bouvieri chevalieri Angel: Loveridge (1947)

Hemidactylus brookii angulatus Hallowell 1852
    Distribution: Cape Verde Islands; Senegal east to Sudan south to
                Tanzania and Angola
    Reference: Lanza and Vanni (1976)
    Hemidactylus bayoni Bocage: Laurent (1964b)

Hemidactylus citernii Boulenger 1912
    Distribution: Somalia and Kenya
    Reference: Loveridge (1957)

Hemidactylus curlei Parker 1942
Distribution: Northern Somalia

Hemidactylus echinus O'Shaughnessy 1875
Distribution: Zaire, Congo, Gabon and Cameroon

Hemidactylus fasciatus fasciatus Gray 1842
Distribution: Liberia east to Zaire

Hemidactylus fasciatus ituriensis Schmidt 1919
Distribution: Eastern Zaire

Hemidactylus flaviviridis Ruppell 1835
Distribution: Egypt south to Somalia; Socotra
Reference: Mahendra (1935a) (1935b) (1936) (1941) (1942); Seshadri (1956)

Hemidactylus forbesii Boulenger 1899
Distribution: Abd el Kuri Island near Socotra

Hemidactylus funaiolii Lanza 1978
Distribution: Central Kenya

Hemidactylus granchii Lanza 1978
Distribution: Eastern central Somalia

Hemidactylus granti Boulenger 1899
Distribution: Socotra

Hemidactylus greeffii Bocage 1886
Distribution: Sao Tome and Principe Islands, Gulf of Guinea

Hemidactylus homoeolepis Blanford 1881
Distribution: Socotra

Hemidactylus isolepis Boulenger 1895
Distribution: Ethiopia, Somalia and Kenya
Reference: Calabresi (1923b) (1927); Lampe (1911); Loveridge (1957)

Hemidactylus jubensis Boulenger 1895
Distribution: Ethiopia and northern Somalia

Hemidactylus laevis Boulenger 1901
Distribution: Somalia

Hemidactylus laticaudatus laticaudatus Andersson 1910
Distribution: Ethiopia
Reference: Calabresi (1925a); Lampe (1911); Scortecci (1928a)

Hemidactylus laticaudatus fossatii Scortecci 1928
Distribution: Eastern Ethiopia

Hemidactylus longicephalus Bocage 1873
Distribution: Angola north to Cameroon
Reference: Loveridge (1964b)

Hemidactylus mabouia mabouia (Moreau de Jonnes 1818)　　　*Gecko*
Distribution: Somalia south to Natal and west to Liberia and Sierra
　　　　　　　Leone; Comoro Islands, Europa Island, Juan de Nova,
　　　　　　　Madagascar, Mascarenes and Seychelles
Reference: Broadley (1977d); Fitzsimons (1943); Lanza and Vanni
　　　　　(1976); Loveridge (1953)
Hemidactylus gardineri Boulenger: Loveridge (1947)
Hemidactylus mercatorius Gray: Loveridge (1953) (1957)

Hemidactylus mabouia tasmani Hewitt 1932
Distribution: Central and eastern Zimbabwe
Reference: Broadley (1977d); Fitzsimons (1943)

Hemidactylus matschiei (Tornier 1901)　　　　　　　　　*Bunocnemis*
Distribution: Nigeria and Togo

Hemidactylus megalops Parker 1932
Distribution: Somalia

Hemidactylus modestus (Gunther 1894)　　　　　　　　　*Bunocnemis*
Distribution: Kenya

Hemidactylus muriceus Peters 1870
Distribution: Guinea east to western Zaire
Reference: Lampe (1911)

Hemidactylus newtoni Ferreira 1897
Distribution: Annobon Island, Gulf of Guinea

Hemidactylus ophiolepis Boulenger 1903
Distribution: Ethiopia and northern Somalia

Hemidactylus ophiolepoides Lanza 1978
Distribution: Northern Somalia

Hemidactylus parkeri Loveridge 1936
Distribution: Kenya north to Ethiopia
Reference: Lanza (1978c)

Hemidactylus platycephalus Peters 1854
Distribution: Somalia south to Mozambique, west to Malawi, eastern
            Zambia and eastern Zimbabwe; Comoro Islands
Reference: Broadley (1977d)

Hemidactylus puccionii Calabresi 1927
Distribution: Somalia and (introduced ?) Zanzibar Island

Hemidactylus pumilio Boulenger 1899
Distribution: Socotra

Hemidactylus richardsonii (Gray 1845)   *Velernesia*
Distribution: Zaire, Congo, Gabon and Cameroon

Hemidactylus ruspolii Boulenger 1896
Distribution: Ethiopia, Somalia and northern Kenya
Reference: Parker (1942); Scortecci (1931b)

Hemidactylus smithi Boulenger 1895
Distribution: Eastern Ethiopia and northern Somalia

Hemidactylus somalicus Parker 1932
Distribution: Northern Somalia

Hemidactylus squamulatus squamulatus Tornier 1896
Distribution: Southern Sudan south through Kenya to central Tanzania
Reference: Calabresi (1923b); Loveridge (1957)
Hemidactylus tropidolepis floweri Werner: Loveridge (1947)

Hemidactylus squamulatus barbouri Loveridge 1942
Distribution: Coastal Kenya and Tanzania

Hemidactylus tanganicus Loveridge 1929
Distribution: Tanzania

Hemidactylus taylori Parker 1932
Distribution: Somalia
Reference: Parker (1942)

Hemidactylus tropidolepis Mocquard 1888
Distribution: Somalia and Kenya
Reference: Andersson (1912), Calabresi (1923b) (1927); Loveridge
(1957)

Hemidactylus turcicus turcicus (Linnaeus 1758)   *Lacerta*
Distribution: Somalia north to Egypt and west to Morocco; Canary
Islands; Socotra.
Reference: Loveridge (1941a); Pasteur and Bons (1960); Roberts and
Schmidt-Nielsen (1966); Tansley (1959)
Hemidactylus sinaitus Boulenger: Parker (1942)

Hemidactylus turcicus macropholis Boulenger 1896
Distribution: Somalia, eastern Ethiopia and northern Kenya
Reference: Scortecci (1931b)

Hemidactylus yerburyi pauciporosus Lanza 1978
Distribution: Northeastern Somalia

Genus: HEMITHECONYX Stejneger 1893
Species typica: *caudicinctus* Dumeril
Reference: Loveridge (1947)

Hemitheconyx caudicinctus (Dumeril 1851)   *Stenodactylus*
Distribution: Nigeria west to Senegal

Hemitheconyx taylori Parker 1930
Distribution: Somalia
Reference: Parker (1942)

Genus: HOLODACTYLUS Boettger 1893
Species typica: *africanus* Boettger
Reference: Loveridge (1947); Parker (1942)

Holodactylus africanus Boettger 1893
Distribution: Eastern Ethiopia and Somalia
Reference: Boulenger (1895g); Calabresi (1918) (1927); Neumann
(1905); Scortecci (1930c) (1931b); Tornier (1905)

Holodactylus cornii Scortecci 1931
    Distribution: Somalis

Genus: HOMOPHOLIS Boulenger 1885
Species typica: *wahlbergii* Smith
Reference: Loveridge (1947)

Homopholis boivini (Dumeril 1856)   *Platydactylus*
    Distribution: Madagascar
    Reference: Angel (1942); Russell (1978)
    Homopholis heterolepis Boulenger: Angel (1942); Loveridge (1947)

Homopholis fasciata fasciata (Boulenger 1890)   *Platypholis*
    Distribution: Kenya and Tanzania

Homopholis fasciata erlangeri Steindachner 1907
    Distribution: Ethiopia and Somalia
    Reference: Scrotecci (1929c)

Homopholis wahlbergii wahlbergii (Smith 1849)   *Gecko*
    Distribution: Southern Mozambique and Transvaal south to Zululand

Homopholis wahlbergii arnoldi Loveridge 1944
    Distribution: Zimbabwe and Botswana

Genus: LYGODACTYLUS Gray 1864
Species typica: *capensis* Smith
Reference: Loveridge (1947)

Lygodactylus angolensis Bocage 1896
    Distribution: Tanzania south to Zimbabwe, west through Botswana,
                Angola and Zaire
    Reference: Broadley (1971e); Loveridge (1957)
    Lygodactylus capensis stevensoni Hewitt: Fitzsimons (1943)

Lygodactylus angularis angularis Gunther 1893
    Distribution: Southwest Tanzania, Malawi and northern Mozambique;
                Zambia
    Reference: Broadley (1971e); Loveridge (1953) (1957)

Lygodatylus angularis heeneni Witte 1933
    Distribution: Eastern Zaire

Lygodactylus arnoulti Pasteur 1964
    Distribution: Madagascar

Lygodactylus blanci Pasteur 1967
    Distribution: Madagascar

Lygodactylus bonsi Pasteur 1962
    Distribution: Malawi

Lygodactylus capensis (Smith 1849)   *Hemidactylus*
    Distribution: Central Kenya south to Cape Province, Namibia, Angola
            and eastern Zaire
    Reference: Brock (1932b); Fitzsimons (1943); Loveridge (1957)
    Lygodactylus capensis bradfieldi Hewitt: Fitzsimons (1943)

Lygodactylus chobiensis Fitzsimons 1932
    Distribution: Northern Botswana and Zambia
    Reference: Broadley (1971e); Fitzsimons (1943)

Lygodactylus conradti Matschie 1892
    Distribution: Eastern Kenya and eastern Tanzania
    Reference: Loveridge (1957)

Lygodactylus conraui Tornier 1902
    Distribution: Cameroon, Gabon and Equatorial Guinea; Liberia
            (introduced ?)

Lygodactylus cowanii (Boulenger 1883)   *Microscalabotes*
    Distribution: Madagascar
    Reference: Angel (1942)

Lygodactylus decaryi Angel 1930
    Distribution: Madagascar
    Reference: Angel (1942)

Lygodactylus expectatus Pasteur and Blanc 1967
    Distribution: Madagascar

Lygodactylus fischeri Boulenger 1890
    Distribution: Sierra Leone, Cameroon, Congo, Gabon and Equatorial
            Guinea

Lygodactylus grandisonae Pasteur 1962
    Distribution: Kenya

Lygodactylus grotei grotei Sternfeld 1911
    Distribution: Tanzania and Mozambique

Lygodactylus grotei pakenhami Loveridge 1941
    Distribution: Pemba Island

Lygodactylus heterurus Boettger 1913
    Distribution: Madagascar
    Reference: Angel (1942)

Lygodactylus insularis Boettger 1913
    Distribution: Juan de Nova Island
    Reference: Angel (1942)

Lygodactylus klemmeri Pasteur 1964
    Distribution: Madagascar

Lygodactylus lawrenci Hewitt 1926
    Distribution: Namibia
    Reference: Fitzsimons (1943)

Lygodactylus madagascariensis (Boettger 1881)   *Scalabotes*
    Distribution: Madagascar
    Reference: Angel (1942)

Lygodactylus methueni Fitzsimons 1937
    Distribution: Transvaal
    Reference: Fitzsimons (1943)

Lygodactylus miops Gunther 1891
    Distribution: Madagascar
    Reference: Angel (1942)

Lygodactylus montanus Pasteur 1964
    Distribution: Madagascar

Lygodactylus ocellatus Roux 1907
    Distribution: Transvaal and Swaziland
    Reference: Fitzsimons (1943)

Lygodactylus ornatus Pasteur 1964
    Distribution: Madagascar

Lygodactylus picturatus picturatus (Peters 1870)   *Hemidactylus*
Distribution: Coastal Kenya, Tanzania, Zambia and southeastern Zaire
Reference: Howell and Mng'ong'o (1978); Loveridge (1957); Pasteur
    (1960b)

Lygodactylus picturatus depressus Schmidt 1919
Distribution: Eastern Zaire

Lygodactylus picturatus gutturalis (Bocage 1873)   *Hemidactylus*
Distribution: Uganda and adjacent Tanzania west to Guinea Bissau

Lygodactylus picturatus keniensis Parker 1936
Distribution: Northern Kenya, southern Somalia and southern Ehtiopia

Lygodactylus picturatus mombasicus Loveridge 1935
Distribution: Coastal southeast Kenya and northeastern Tanzania

Lygodactylus picturatus sudanensis Loveridge 1935
Distribution: Ehtiopia and southern Sudan

Lygodactylus picturatus ukerewensis Loveridge 1935
Distribution: Kenyan and Tanzanian banks of Lake Victoria

Lygodactylus picturatus williamsi Loveridge 1952
Distribution: Eastern Province of Tanzania

Lygodactylus robustus Boettger 1913
Distribution: Madagascar
Reference: Angel (1942)

Lygodactylus scheffleri Sternfeld 1912
Distribution: Kenya

Lygodactylus septemtuberculatus Angel 1942
Distribution: Madagascar

Lygodactylus somalicus somalicus Loveridge 1935
Distribution: Somalia
Reference: Loveridge (1957); Parker (1942)
Lygodactylus somalicus annectens Loveridge: Loveridge (1947)

Lygodactylus somalicus battersbyi Pasteur 1962
Distribution: Northern Kenya and adjacent Somalia

Lygodactylus spinulifer (Boettger 1913)   *Microscalabotes*
    Distribution: Madagascar
    Reference: Angel (1942)

Lygodactylus thomensis thomensis (Peters 1880)   *Scalabotes*
    Distribution: Sao Tome Island, Gulf of Guinea

Lygodactylus thomensis delicatus Pasteur 1962
    Distribution: Principe Island, Gulf of Guinea

Lygodactylus thomensis wermuthi Pasteur 1962
    Distribution: Annobon Island, Gulf of Guinea

Lygodactylus tolampyae (Grandidier 1872)   *Hemidactylus*
    Distribution: Madagascar
    Reference: Angel (1942)

Lygodactylus tuberifer Boettger 1913
    Distribution: Madagascar
    Reference: Angel (1942)

Lygodactylus tuberosus Mertens 1965
    Distribution: Madagascar

Lygodactylus verticillatus Mocquard 1895
    Distribution: Madagascar
    Reference: Angel (1942)

Genus: MEDIODACTYLUS Szczerbak and Golubev 1977
Species typica: *kotschyi* Steindachner

Mediodactylus kotschyi (Steindachner 1870)   *Gymnodactylus*
    Distribution: Egypt
    Reference: Arnold and Burton (1978); Loveridge (1947); Marx (1968)

Genus: NARUDASIA Methuen and Hewitt 1914
Species typica: *festiva* Methuen and Hewitt

Narudasia festiva Methuen and Hewitt 1914
    Distribution: Namibia
    Reference: Fitzsimons (1943); Loveridge (1947)

Genus: PACHYDACTYLUS Wiegmann 1834
Species typica: *geitje* Sparrman
Reference: Fitzsimons (1943); Loveridge (1947)

Pachydactylus amoenus Werner 1910
  Distribution: Cape Province

Pachydactylus austeni Hewitt 1923
  Distribution: Cape Province

Pachydactylus barnardi Fitzsimons 1941
  Distribution: Cape Province
  Reference: Broadley (1977c)

Pachydactylus bibroni bibroni (Smith 1845)   *Tarentola*
  Distribution: Cape Province

Pachydactylus bibroni pulitzerae Schmidt 1933
  Distribution: Namibia and southern Angola
  Reference: Laurent (1964b)

Pachydactylus bibroni turneri (Gray 1864)   *Homodactylus*
  Distribution: Tanzania, Rwanda, Burundi, Mozambique, Malawi,
            Zambia, Zimbabwe, Orange Free State, Transvaal,
            Botswana and Namibia

Pachydactylus bicolor Hewitt 1926
  Distribution: Namibia

Pachydactylus capensis capensis (Smith 1845)   *Tarentola*
  Distribution: Southern Angola south through Botswana to northern
            Transvaal and central Cape Province
  Reference: Broadley (1977c)
  Pachydactylus mentalis Hewitt: Fitzsimons (1943)

Pachydactylus capensis affinis Boulenger 1896
  Distribution: Transvaal

Pachydactylus capensis formosus Smith 1849
  Distribution: Southern Transvaal

Pachydactylus capensis katanganus Witte 1953
  Distribution: Shaba Province, Zaire
  Reference: Broadley (1971e) (1977c)

Pachydactylus capensis labialis Fitzsimons 1938
    Distribution: Cape Province

Pachydactylus capensis oshaughnessyi Boulenger 1885
    Distribution: Northern Zimbabwe and adjacent western Mozambique,
            southern Malawi and Zambia
    Reference: Broadley (1971e) (1977c)
    Pachydactylus capensis levyi Fitzsimons: Loveridge (1947)

Pachydactylus capensis vansoni Fitzsimons 1933
    Distribution: Northeast and eastern Transvaal, southeast Zimbabwe
            and southern Mozambique
    Reference: Broadley (1977c)

Pachydactylus fasciatus Boulenger 1888
    Distribution: Namibia
    Reference: Haacke (1965)

Pachydactylus geitje (Sparrman 1778)   *Lacerta*
    Distribution: Cape Province
    Reference: Bustard (1963b)
    Pachydactylus ocellatus (Cuvier): Fitzsimons (1943)

Pachydactylus kobosensis Fitzsimons 1938
    Distribution: Namibia

Pachydactylus laevigatus laevigatus Fischer 1888
    Distribution: Southern Angola south to northwest Cape Province
    Reference: Werner (1977)

Pachydactylus laevigatus fitzsimonsi Loveridge 1947
    Distribution: Namibia and southern Angola
    Reference: Laurent (1964b)
    Pachydactylus laevigatus tessellatus Fitzsimons: Fitzsimons (1943)

Pachydactylus maculatus Gray 1845
    Distribution: Zululand and Natal west through Cape Province
    Reference: Brock (1932b)
    Pachydactylus capensis oculatus Hewitt: Fitzsimons (1943)
    Pachydactylus maculatus albomarginatus Hewitt: Fitzsimons (1943)
    Pachydactylus microlepis Hewitt: Fitzsimons (1943)

Pachydactylus mariquensis mariquensis Smith 1849
    Distribution: Southern Orange Free State south to southern Cape
            Province

Pachydactylus mariquensis latirostris Hewitt 1923
Distribution: Northwestern Cape Province and southern Namibia

Pachydactylus mariquensis macrolepis Fitzsimons 1939
Distribution: Cape Province

Pachydactylus monticolus Fitzsimons 1943
Distribution: Cape Province

Pachydactylus namaquensis (Sclater 1898)   *Elasmodactylus*
Distribution: Western Cape Province and southern Namibia

Pachydactylus oreophilus oreophilus McLachlan and Spence 1967
Distribution: Namibia

Pachydactylus oreophilus gaiasensis Steyn and Mitchell 1967
Distribution: Namibia

Pachydactylus punctatus punctatus Peters 1854
Distribution: Mozambique west through Transvaal, Botswana, Zambia,
            Namibia and Angola
Reference: Broadley (1971e); Laurent (1964b)

Pachydactylus punctatus amoenoides Hewitt 1935
Distribution: Namibia
Reference: Laurent (1964b); Werner (1977)

Pachydactylus rugosus rugosus Smith 1849
Distribution: Northwestern Cape Province and southern Namibia

Pachydactylus rugosus frater Hewitt 1935
Distribution: Northern Namibia

Pachydactylus scutatus scutatus Hewitt 1927
Distribution: Northern Namibia

Pachydactylus scutatus angolensis Loveridge 1944
Distribution: Southern Angola
Reference: Laurent (1964b)

Pachydactylus scutatus robertsi Fitzsimons 1938
Distribution: Southern Namibia

Pachydactylus serval serval Werner 1910
   Distribution: Namibia and southern Angola
   Reference: McLachlan and Spence (1966b)
   Pachydactylus montanus montanus Methuen and Hewitt: Fitzsimons
                        (1943)

Pachydactylus serval onscepensis Hewitt 1935
   Distribution: Namibia

Pachydactylus serval purcelli Boulenger 1910
   Distribution: Western Cape Province and southern Namibia

Pachydactylus serval sansteyni Steyn and Mitchell 1967
   Distribution: Namibia

Pachydactylus tetensis Loveridge 1953
   Distribution: Southern Tanzania and Mozambique

Pachydactylus tigrinus Van Dam 1821
   Distribution: Northern Transvaal, Zimbabwe, eastern Botswana and
                        west Mozambique
   Reference: Broadley (1977c)
   Pachydactylus capensis rhodesianus Loveridge: Loveridge (1947)

Pachydactylus tuberculosus (Boulenger 1894)   *Elasmodactylus*
   Distribution: Tanzania, Zaire and Zambia

Pachydactylus weberi weberi Roux 1907
   Distribution: Cape Province

Pachydactylus weberi acuminatus Fitzsimons 1941
   Distribution: Southern Namibia

Pachydactylus weberi gariesensis Hewitt 1932
   Distribution: Cape Province

Pachydactylus weberi werneri Hewitt 1935
   Distribution: Coastal central Namibia

Genus: PARAGEHYRA Angel 1929
Species typica: *petiti* Angel

Paragehyra petiti Angel 1929
    Distribution: Madagascar
    Reference: Angel (1942)

Genus: PHYLLODACTYLUS Gray 1828
Species typica: *pulcher*
Reference: Fitzsimons (1943); Loveridge (1947)

Phyllodactylus ansorgii Boulenger 1907
    Distribution: Angola

Phyllodactylus brevipes Mocquard 1900
    Distribution: Madagascar
    Reference: Angel (1942)

Phyllodactylus europaeus Gene 1839
    Distribution: Galaite Island, Tunisia
    Reference: Arnold and Burton (1978); Pellegrin (1927a); Wiedersheim
        (1876)

Phyllodactylus inexpectatus (Stejneger 1893)   *Diplodactylus*
    Distribution: Seychelles
    Reference: Loveridge (1947)

Phyllodactylus lineatus lineatus Gray 1838
    Distribution: Western Cape Province
    Reference: Haacke (1965)

Phyllodactylus lineatus essexi Hewitt 1925
    Distribution: Eastern Cape Province

Phyllodactylus lineatus rupicolus Fitzsimons 1938
    Distribution: Cape Province

Phyllodactylus microlepidotus Fitzsimons 1939
    Distribution: Cape Province

Phyllodactylus palmatus Mocquard 1902
    Distribution: Congo and Cameroon

Phyllodactylus peringueyi Boulenger 1910
   Distribution: Cape Province
   Comment: Both Fitzsimons (1943) and Loveridge (1947) comment on
            the non-African relation of this gecko, its affinities being
            with American species (introduced ?)

Phyllodactylus porphyreus porphyreus (Daudin 1802)   *Gecko*
   Distribution: Cape Province
   Reference:
   Phyllodactylus porphyreus cronwrighti Hewitt: Fitzsimons (1943)

Phyllodactylus porphyreus namaquensis Hewitt 1935
   Distribution: Cape Province

Phyllodactylus riebeckii (Peters 1882)   *Diplodactylus*
   Distribution: Socotra

Phyllodactylus trachyrhinus Boulenger 1899
   Distribution: Socotra

Phyllodactylus weileri (Muller 1909)   *Diplodactylus*
   Distribution: Cameroon

Phyllodactylus wolterstorffi (Tornier 1900)   *Diplodactylus*
   Distribution: Tanzania

Genus: PRISTURUS Ruppell 1835
Species typica: *flavipunctatus* Ruppell
Reference: Loveridge (1947); Parker (1942); Scortecci (1935c)

Pristurus crucifer (Valenciennes 1861)   *Gymnocephalus*
   Distribution: Ehtiopia, Somalia and Socotra
   Reference: Tornier (1905)

Pristurus flavipunctatus flavipunctatus Ruppell 1835
   Distribution: Southeastern Egypt south to Somalia

Pristurus insignis Blanford 1881
   Distribution: Socotra

Pristurus phillipsii Boulenger 1895
   Distribution: Somalia
   Reference: Calabresi (1915) (1927)

Pristurus rupestris rupestris Blanford 1874
    Distribution: Somalia

Pristurus sokotranus Parker 1938
    Distribution: Socotra

Genus: PTYODACTYLUS Goldfuss 1820
Species typica: *hasselquistii* Donndorff
Reference: Loveridge (1947)

Ptyodactylus hasselquistii hasselquistii (Donndorff 1798)   *Lacerta*
    Distribution: Egypt and Libya west through southern Algeria and
                extreme southern Morocco to Western Sahara and
                Mauritania
    Reference: Tercafs (1962)

Ptyodactylus hasselquistii oudrii Lataste 1880
    Distribution: Northern Algeria and Morocco
    Reference: Pasteur and Bons (1960)

Ptyodactylus hasselquistii sokotranus Steindachner 1902
    Distribution: Socotra

Ptyodactylus hasselquistii togoensis Tornier 1901
    Distribution: Northern Togo

Genus: STENODACTYLUS Fitzinger 1826
Species typica: *sthenodactylus* Lichtenstein
Reference: Loveridge (1947); Pasteur and Bons (1960); Scortecci (1935d)

Stenodactylus petrii Anderson 1896
    Distribution: Egypt west to extreme eastern Morocco
    Reference: Lampe (1911); Marx (1968)

Stenodactylus sthenodactylus sthenodactylus (Lichtenstein 1823)
                                                        Ascalabotes
    Distribution: Tunisia east to Egypt then south to northern Kenya
    Reference: Loveridge (1957); Werner (1964)

Stenodactylus sthenodactylus mauritanicus Guichenot 1850
    Distribution: Algeria, Saharan Morocco, Western Sahara and Mauritania
    Reference: Doumergue (1899); Pellegrin (1910) (1926a) (1926f) (1927a)

Genus: TARENTOLA Gray 1825
Species typica: *mauritanica* Linnaeus
Reference: Loveridge (1947)

Tarentola annularis (Geoffroy 1827)   *Gecko*
  Distribution: Egypt, Libya, Sudan, Ethiopia and northern Somalia west
              through Chad, Niger and Mali avoiding coastal West Africa
              to Gambia, Senegal and Guinea Bissau; Mauritania,
              Western Sahara and southern Morocco
  Reference: Cloudsley-Thompson (1972); Grandison (1961); Hoofien
              (1962)

Tarentola delalandii delalandii (Dumeril and Bibron 1836)
                                                       Platydactylus
  Distribution: Madeira, Canary and Cape Verde Islands
  Reference:
  Tarentola delalandii boettgeri Steindachner: Loveridge (1947)

Tarentola delalandii gigas (Bocage 1875)   *Ascalabotes*
  Distribution: Branco and Raso, Cape Verde Islands

Tarentola ephippiata O'Shaughnessy 1875
  Distribution: Morocco south to Senegal, Gambia and Guinea Bissau,
              east avoiding the West African coast to northern
              Cameroon, Chad and northern Sudan
  Reference: Grandison (1961)
  Tarentola delalandii hoggarensis Werner: Loveridge (1947)
  Tarentola hoggarensis panousei Pasteur: Pasteur and Bons (1960)

Tarentola mauritanica (Linnaeus 1758)   *Lacerta*
  Distribution: Egypt west to Morocco and Western Sahara; Canary
              Islands and Madeira
  Reference: Arnold and Burton (1978); Gabe and Saint Girons (1969);
              Hiller (1977); Pasteur and Girot (1960); Tansley (1959);
              Wood (1938)
  Tarentola mauritanica deserti Boulenger: Loveridge (1947)

Tarentola neglecta Strauch 1887
  Distribution: Algeria, Tunisia and Libya

Genus: TROPIOCOLOTES Peters 1880
Species typica: *tripolitanus* Peters
Reference: Leviton and Anderson (1972); Loveridge (1947); Minton,
          Anderson and Anderson (1970); Parker (1942)

Tropiocolotes nattereri Steindachner 1901
    Distribution: Egypt and eastern Libya
    Reference: Pasteur (1960c)

Tropiocolotes steudneri (Peters 1869)   *Gymnodactylus*
    Distribution: Egypt and Sudanwest to Algerian Sahara

Tropiocolotes tripolitanus tripolitanus Peters 1880
    Distribution: Egypt west to Tunisia

Tropiocolotes tripolitanus algericus Loveridge 1947
    Distribution: Algerian and Moroccan Sahara

Tropiocolotes tripolitanus apoklomax Papenfuss 1969
    Distribution: Mali

Tropiocolotes tripolitanus occidentalis Parker 1942
    Distribution: Western Sahara

Tropiocolotes tripolitanus somalicus Parker 1942
    Distribution: Somalia

Genus: UROPLATUS Dumeril 1805
Species typica: *fimbriatus* Schneider
Reference: Angel (1942)

Uroplatus alluaudi Mocquard 1894
    Distribution: Madagascar

Uroplatus ebenaui Boettger 1879
    Distribution: Madagascar

Uroplatus fimbriatus (Schneider 1797)   *Stellio*
    Distribution: Madagascar

Uroplatus guentheri Mocquard 1908
    Distribution: Madagascar

Uroplatus lineatus (Dumeril and Bibron 1836)   *Ptyodactylus*
    Distribution: Madagascar

Uroplatus phantasticus Boulenger 1888
    Distribution: Madagascar
    Reference: Angel (1930a)

Chapter 5

FAMILY IGUANIDAE

Order: Sauria
Suborder: Ascalabota
Infraorder: Iguania
Family: Iguanidae

Genus: CHALARODON Peters 1854
Species typica: *madagascariensis* Peters

Chalarodon madagascariensis Peters 1854
    Distribution: Madagascar
    Reference: Avery and Tanner (1971); Blanc (1965) (1969); Blanc and
        Carpenter (1969); Petit (1928)

Genus: OPLURUS Cuvier 1829
Species typica: *cyclurus* Merrem
Reference: Angel (1942)

Oplurus cyclurus (Merrem 1820)   *Uromastix*
    Distribution: Madagascar

Oplurus fierinensis Grandidier 1869
    Distribution: Madagascar

Oplurus grandidieri Mocquard 1900
    Distribution: Madagascar

Oplurus quadrimaculatus Dumeril 1851
    Distribution: Madagascar

Oplurus saxicola Grandidier 1869
    Distribution: Madagascar

Oplurus sebae sebae Dumeril and Bibron 1837
    Distribution: Madagascar

Oplurus sebae comorensis Angel 1942
    Distribution: Comoro Islands

Chapter 6

FAMILY AGAMIDAE

Order: Sauria
Suborder: Ascalabota
Infraorder: Iguania
Family: Agamidae

Genus: AGAMA Daudin 1802
Species typica: *agama* Linnaeus

Agama agama (Linnaeus 1758)   *Lacerta*
    Distribution: Senegal east to Ethiopia and south to Tanzania in the east
            and Angola in the west
    Reference: Bradley (1903); Chapman and Chapman (1964); Charnier
            (1965); Grandison (1968); Halstead (1970); James and Porter
            (1979); Porter and James (1979); Toye (1972)
    Agama agama africana (Hallowell); Daniel (1961)
    Agama agama dodomae Loveridge: Loveridge (1957)
    Agama agama elgonis Lonnberg: Loveridge (1957)
    Agama agama lionotus Boulenger: Loveridge (1957); Marshall and
            Hook (1960)
    Agama agama savattieri Rochebrune: Grandison (1956)
    Agama agama ufipae Loveridge: Loveridge (1957)
    Agama agama usambarae Barbour and Loveridge: Loveridge (1957)

Agama anchietae Bocage 1896
    Distribution: Cape Province north through Namibia, Angola and
            southern Zaire
    Reference: Fitzsimons (1943); Laurent (1964b)

Agama atra atra Daudin 1802
   Distribution: South Africa, southern Namibia and southern Botswana
   Reference: Bruton (1977); Burrage (1974); Fitzsimons (1943)

Agama atra knobeli Boulenger and Power 1921
   Distribution: Namibia
   Reference: Fitzsimons (1943)

Agama atricollis atricollis Smith 1849
   Distribution: South Africa north to Tanzania in the east and Angola in
         the west
   Reference: Cowles (1956); Fitzsimons (1943); Schmidt (1966); Schmidt
         and Inger (1957)

Agama atricollis gregorii Gunther 1894
   Distribution: Coastal Kenya and adjacent Tanzania

Agama atricollis kiwuensis Klausewitz 1957
   Distribution: Zaire

Agama atricollis loveridgei Klausewitz 1957
   Distribution: Zambia north through Tanzania to Sudan

Agama atricollis minuta Klausewitz 1957
   Distribution: Ehtiopia and Kenya

Agama atricollis ugandaensis Klausewitz 1957
   Distribution: Uganda and adjacent Tanzania

Agama batillifera (Vaillant 1882)   *Uromastix*
   Distribution: Somalia
   Reference: Parker (1942)

Agama benueensis Monard 1951
   Distribution: Northern Nigeria and northern Cameroon
   Reference: Grandison (1968)

Agama bocourti Rochebrune 1884
   Distribution: Senegal and Gambia

Agama bottegi Boulenger 1898
   Distribution: Somalia
   Reference: Lanza (1978a)

Agama boueti Chabanaud 1917
    Distribution: Mali

Agama boulengeri Lataste 1886
    Distribution: Senegal

Agama caudospinosa Meek 1910
    Distribution: Kenya

Agama cornii Scortecci 1928
    Distribution: Ethiopia

Agama cristata Mocquard 1905
    Distribution: Mali

Agama cyanogaster cyanogaster (Ruppell 1835)   *Stellio*
    Distribution: Ethiopia
    Reference: Curry-Lindahl (1957); Klausewitz (1954); Loveridge (1957);
            Robertson and others (1965)

Agama cyanogaster annectens Blanford 1870
    Distribution: Ethiopia, Somalis and Kenya
    Reference: Klausewitz (1954); Loveridge (1957)

Agama cyanogaster phillipsii Boulenger 1895
    Distribution: Coastal Somalia and Ethiopia
    Reference: Klausewitz (1954)

Agama flavimaculata (Ruppell 1835)   *Trapelus*
    Distribution: Egypt
    Reference: Pasteur and Bons (1960)

Agama gracilimembris Chabanaud 1918
    Distribution: Benin east to Central African Republic
    Reference: Grandison (1968)

Agama hartmanni Peters 1869
    Distribution: Southern Egypt, Sudan and adjacent Ethiopia

Agama hispida hispida (Linnaeus 1754)   *Lacerta*
    Distribution: Western and southwestern Cape Province
    Reference: Barry (1953); Fitzsimons (1943); Huey and Pianka (1977)

Agama hispida aculeata Merrem 1820
Distribution: Northern Cape Province, Zimbabwe, Namibia and
southern Angola
Reference: Fitzsimons (1943); Laurent (1964b)

Agama hispida brachyura Boulenger 1885
Distribution: Northwestern Cape Province and southern Namibia
Reference: Fitzsimons (1943)

Agama hispida distanti Boulenger 1902
Distribution: Transvaal, Zimbabwe and adjacent Botswana
Reference: Fitzsimons (1943)

Agama hispida makarikarica Fitzsimons 1932
Distribution: Botswana
Reference: Fitzsimons (1943); Steyn, Kinkeldey and Buys (1963)

Agama hispida mertensi Wermuth 1967
Distribution: Tanzania, Mozambique, Malawi, Zimbabwe, Swaziland
and adjacent South Africa
Reference:
Agama hispida armata Peters: Fitzsimons (1943); Loveridge (1957)

Agama impalearis Boettger 1874
Distribution: Morocco, Algeria and Tunisia
Reference:
Agama bibronii Dumeril: Bons (1964) (1968b); Pasteur and Bons
(1960); Saint Girons (1967)

Agama insularis Chabanaud 1918
Distribution: Rooma Island, Las Islands off the coast of Guinea

Agama kirkii Boulenger 1885
Distribution: Tanzania, Malawi, Zambia, Zimbabwe, Mozambique and
Botswana
Reference: Fitzsimons (1943)

Agama mehelyi Tornier 1902
Distribution: Cameroon

Agama mossambica mossambica Peters 1854
Distribution: Tanzania, Malawi, Zambia and Mozambique
Reference: Fitzsimons (1943); Loveridge (1957)

Agama mossambica montana Barbour and Loveridge 1928
    Distribution: Uluguru and Usambara Mountains, Tanzania
    Reference: Loveridge (1957)

Agama mutabilis Merrem 1820
    Distribution: Saharan Morocco east to Egypt
    Reference: Pasteur and Bons (1960)

Agama paragama Grandison 1968
    Distribution: Northern Nigeria and northern Cameroon

Agama persimilis Parker 1942
    Distribution: Somalia
    Reference: Lanza (1978a)

Agama planiceps planiceps Peters 1862
    Distribution: Namibia
    Reference: Fitzsimons (1943)

Agama planiceps mwanzae Loveridge 1923
    Distribution: Northwestern Tanzania
    Reference: Loveridge (1957)

Agama planiceps schacki Mertens 1938
    Distribution: Angola
    Reference: Fitzsimons (1943); Laurent (1943b)

Agama robecchii Boulenger 1892
    Distribution: Northern Somalia
    Reference: Parker (1942)

Agama rueppelli rueppelli Vaillant 1882
    Distribution: Somalia and adjacent Ehtiopia

Agama rueppelli occidentalis Parker 1932
    Distribution: Southern Ethiopia and northern Kenya
    Reference: Loveridge (1957)

Agama rueppelli septentrionalis Parker 1932
    Distribution: Central and southern Kenya
    Reference: Loveridge (1957)

Agama sankaranica Chabanaud 1918
    Distribution: Guinea east to Nigeria
    Reference: Grandison (1968)

Agama savignii Dumeril and Bibron 1837
  Distribution: Egypt
  Reference: Marx (1968)

Agama sennariensis Werner 1914
  Distribution: Sudan

Agama sinaita Heyden 1827
  Distribution: Egypt south to Ethiopia east of the Nile

Agama spinosa Gray 1831
  Distribution: Egypt south to northern Ethiopia
  Reference: Lanza (1978a); Parker (1942)

Agama stellio vulgaris (Sonnini and Latreille 1802)   *Stellio*
  Distribution: Northern Egypt

Agama taylori Parker 1935
  Distribution: Somalia
  Reference: Parker (1942)

Agama tournevillei Lataste 1880
  Distribution: Algerian Sahara
  Reference: Pasteur and Bons (1960)

Agama weidholzi Wettstein 1932
  Distribution: Senegal and Gambia

Agama zonura Boulenger 1895
  Distribution: Western Somalia

Genus: UROMASTYX Merrem 1820
Species typica: *aegyptius* Forskal
Reference: Mertens (1962); Pasteur and Bons (1960)

Uromastyx acanthinurus acanthinurus Bell 1825
  Distribution: Northern Algeria and Tunisia east to Egypt
  Reference: Bentley (1976); Grenot (1967) (1968a) (1968b) (1974);
             Komnick (1970); Lemire, Deloince and Grenot (1970)
             (1972); Schmidt and Inger (1957); VanLennep and Komnick
             (1970); Vogel (1964)

Uromastyx acanthinurus dispar Heyden 1827
  Distribution: Northern Sudan

Uromastyx acanthinurus flavifasciatus Mertens 1962
  Distribution: Senegal, Mauritania and Western Sahara

Uromastyx acanthinurus geyri Muller 1922
  Distribution: The highlands of the central Sahara

Uromastyx acanthinurus nigerrimus Hartert 1913
  Distribution: Saharan Algeria

Uromastyx acanthinurus werneri Muller 1922
  Distribution: Western Algeria and Morocco

Uromastyx aegyptius (Forskal 1775)   *Lacerta*
  Distribution: Egypt
  Reference: El-Toubi and Bishai (1959); Khalil and Abdel-Messeih (1954)
         (1961a); Khalil and Hussein (1962) (1963); Khalil and Yanni
         (1959) (1961); Schmidt-Nielsen and others (1963); Sokolov
         (1966); Throckmorton (1976)

Uromastyx macfadyeni Parker 1932
  Distribution: Northern Somalia
  Reference: Parker (1942)

Uromastyx ocellatus Lichtenstein 1823
  Distribution: Egypt

Uromastyx ornatus Heyden 1827
  Distribution: Northeastern Egypt

Uromastyx princeps O'Shaughnessy 1880
  Distribution: Ethiopia, Somalia, coastal Kenya and Zanzibar
  Reference: Parker (1942)

Chapter 7

FAMILY CHAMAELEONIDAE

Order: Sauria
Suborder: Ascalabota
Infraorder: Iguania
Family: Chamaeleonidae

Genus: BRADYPODION Fitzinger 1843
Species typica: *pumila* Gmelin
Reference: Fitzsimons (1943); Raw (1976)

Bradypodion caffrum (Boettger 1889)   *Chamaeleo*
    Distribution: vicinity of Port St. Johns, eastern South Africa

Bradypodion damaranum (Boulenger 1887)   *Chamaeleon*
    Distribution: Southern Cape Province
    Reference: Spence (1966)

Bradypodion dracomontanum Raw 1976
    Distribution: Southern and central Natal Drakensberg

Bradypodion gutturale (Smith 1849)   *Chamaeleo*
    Distribution: South Africa

Bradypodion karroicum (Methuen and Hewitt 1915)   *Lophosaura*
    Distribution: Southern Cape Province

Bradypodion melanocephalum (Gray 1864)   *Microsaura*
    Distribution: Coastal Natal

Bradypodion nemorale Raw 1978
   Distribution: Qudeni and Nkandla forests, Zululand

Bradypodion occidentale (Hewitt 1935)   *Lophosaura*
   Distribution: Northwestern Cape Province and southern Namibia

Bradypodion pumilum (Gmelin 1789)   *Lacerta*
   Distribution: Western Cape Province
   Reference: Atsatt (1953); Burrage (1973); Englebrecht (1951); Frisch
          (1962); Schmidt and Inger (1957); Skinner (1959); Visser
          (1972)

Bradypodion setaroi Raw 1976
   Distribution: Coastal dune forest of Zululand

Bradypodion taeniabronchum (Smith 1831)   *Chamaeleo*
   Distribution: vicinity of Algoa Bay, Cape Province

Bradypodion thamnobates Raw 1976
   Distribution: Natal Midlands

Bradypodion transvaalense (Fitzsimons 1930)   *Chamaeleon*
   Distribution: Eastern South Africa

Bradypodion ventrale (Gray 1845)   *Chamaeleo*
   Distribution: Eastern Cape Province
   Reference: Brock (1940)

Genus: BROOKESIA Gray 1864
Species typica: *superciliaris* Kuhl
Reference: Angel (1942)

Brookesia decaryi Angel 1938
   Distribution: Madagascar

Brookesia dentata Mocquard 1900
   Distribution: Madagascar

Brookesia ebenaui (Boettger 1880)   *Chamaeleon*
   Distribution: Madagascar

Brookesia minima Boettger 1893
   Distribution: Madagascar

Brookesia nasus Boulenger 1887
   Distribution: Madagascar

Brookesia perarmata (Angel 1933)   *Leandria*
   Distribution: Madagascar

Brookesia stumpffi Boettger 1894
   Distribution: Madagascar
   Reference: Mertens (1951)

Brookesia superciliaris (Kuhl 1820)   *Chamaeleon*
   Distribution: Madagascar

Brookesia tuberculata Mocquard 1894
   Distribution: Madagascar

Genus: CHAMAELEO Laurenti 1768
Species typica: *chamaeleon* Linnaeus
Reference: Angel (1942); Hillenius (1959) (1963); Loveridge (1957)

Chamaeleo adolfifriderici Sternfeld 1912
   Distribution: Zaire, Rwanda and Uganda

Chamaeleo affinis Ruppell 1845
   Distribution: Ethiopia and adjacent Somalia

Chamaeleo africanus Laurenti 1768
   Distribution: Nigeria and Cameroon east to Somalia, Ethiopia and
             Sudan
   Reference: Gabe and Saint Girons (1965)

Chamaeleo anchietae anchietae Bocage 1872
   Distribution: Angola

Chamaeleo anchietae vinckei Laurent 1950
   Distribution: Southern Zaire and southern Tanzania

Chamaeleo antimena Grandidier 1872
   Distribution: Madagascar

Chamaeleo bitaeniatus Fischer 1884
   Distribution: Ethiopia, Somalia, Uganda, Kenya, north and west Tanzania
   Reference: Bustard (1966); Rand (1963); Sjongren (1945)

Chamaeleo boettgeri Boulenger 1888
Distribution: Madagascar

Chamaeleo brevicornis Gunther 1879
Distribution: Madagascar

Chamaeleo campani Grandidier 1872
Distribution: Madagascar

Chamaeleo cephalolepis Gunther 1880
Distribution: Comoro Islands

Chamaeleo chamaeleon chamaeleon (Linnaeus 1758)   *Lacerta*
Distribution: North coast of Africa
Reference: Arnold and Burton (1978); Bons and Bons (1960); Grunwald
    (1931); Haas (1937); Horn (1947); Pasteur and Bons (1960)

Chamaeleo chapini Witte 1964
Distribution: Zaire

Chamaeleo cristatus Stutchbury 1837
Distribution: Nigeria, Cameroon, Gabon and Equatorial Guinea

Chamaeleo cucullatus Gray 1831
Distribution: Madagascar

Chamaeleo deremensis Matschie 1892
Distribution: Usambara and Uluguru Mountains, Tanzania
Reference: Rand (1958)

Chamaeleo dilepis dilepis Leach 1819
Distribution: Cameroon east to Kenya and south to northern Namibia,
    Botswana and northeastern South Africa
Reference: Brain (1961); Fitzsimons (1943); Laurent (1964b); Schmidt
    and Inger (1957); Vogel (1964)

Chamaeleo dilepis idjwiensis Loveridge 1942
Distribution: Rwanda, Burundi and adjacent Zaire

Chamaeleo dilepis isabellinus Gunther 1893
Distribution: Shire Highlands, Malawi

Chamaeleo dilepis petersii Gray 1864
Distribution: Mozambique and southern Tanzania

Chamaeleo dilepis roperi Boulenger 1890
Distribution: Eastern Kenya and northeastern Tanzania

Chamaeleo dilepis ruspolii Boettger 1893
Distribution: Somalia
Reference: Parker (1942)

Chamaeleo ellioti Gunther 1895
Distribution: Sudan, Kenya, Tanzania, Uganda, Rwanda, Burundi and
            eastern Zaire
Reference: Herlant and Pasteels (1955); Rand (1963)

Chamaeleo fallax Mocquard 1900
Distribution: Madagascar

Chamaeleo fischeri fischeri Reichenow 1887
Distribution: Tanzania: eastern Usambara Mountains south to Nguru
            Mountains
Reference: Schmidt and Inger (1957)

Chamaeleo fischeri excubitor Barbour 1911
Distribution: Mount Kenya, Kenya

Chamaeleo fischeri multituberculatus Nieden 1913
Distribution: Western Usambara Mountains, Tanzania

Chamaeleo fischeri tavetanus Steindachner 1891
Distribution: Teita Mountains of Kenya west to Kilimanjaro, Tanzania

Chamaeleo fischeri uluguruensis Loveridge 1956
Distribution: Uluguru Mountains, Tanzania

Chamaeleo fischeri uthmolleri Muller 1938
Distribution: Mount Hanang, central Tanzania

Chamaeleo fuelleborni Tornier 1900
Distribution: Ngosi Volcano, Poroto Mountains, Tanzania
Reference: Rand (1958)

Chamaeleo furcifer Vaillant and Grandidier 1880
Distribution: Madagascar

Chamaeleo gallus Gunther 1877
Distribution: Madagascar

Chamaeleo gastrotaenia Boulenger 1888
Distribution: Madagascar

Chamaeleo globifer Gunther 1879
Distribution: Madagascar

Chamaeleo goetzei goetzei Tornier 1899
Distribution: Tanzania

Chamaeleo goetzei nyikae Loveridge 1953
Distribution: Nyika Plateau, Malawi

Chamaeleo gracilis gracilis Hallowell 1842
Distribution: Somalia, Kenya, northern Tanzania west through Zaire,
Uganda, and Central African Republic to Senegal and
Gambia
Reference: Bustard (1967); Lanza and Vanni (1976); Menzies (1958)

Chamaeleo gracilis etiennei Schmidt 1919
Distribution: Gabon south to Angola
Reference: Laurent (1964b)

Chamaeleo guibei Hillenius 1959
Distribution: Madagascar

Chamaeleo hohnelii Steindachner 1891
Distribution: Kenya and Uganda
Reference: Bustard (1965); Lin (1979); Rand (1963)
Chamaeleo bitaeniatus altaeelgonis Loveridge: Loveridge (1957)

Chamaeleo incornutus Loveridge 1932
Distribution: Ukinga, Rungwe and Poroto Mountains, Tanzania

Chamaeleo jacksonii jacksonii Boulenger 1896
Distribution: Kenya and Tanzania
Reference: Bustard (1958) (1963); Lin (1979); Rand (1958); Schmidt and
Inger (1957)

Chamaeleo jacksonii merumontanus Rand 1958
Distribution: Mount Meru, Tanzania

Chamaeleo johnstoni johnstoni Boulenger 1901
Distribution: Uganda, Rwanda, Burundi and adjacent Zaire
Reference: Rand (1958)
Chamaeleo johnstoni crenulatus Laurent: Rand (1958)

Chamaeleo johnstoni ituriensis Schmidt 1919
  Distribution: Eastern Zaire
  Reference: Rand (1958)

Chamaeleo kinetensis Schmidt 1943
  Distribution: Imatong Mountains, Sudan
  Reference: Rand (1963)

Chamaeleo lateralis Gray 1831
  Distribution: Madagascar
  Reference:
  Chamaeleo lambertoni Angel: Angel (1942)

Chamaeleo laterispinis Loveridge 1932
  Distribution: Uzungwe Mountains, Tanzania

Chamaeleo linotus Muller 1924
  Distribution: Madagascar

Chamaeleo malthe Gunther 1879
  Distribution: Madagascar

Chamaeleo melleri (Gray 1864)   *Ensirostris*
  Distribution: Malawi and Tanzania

Chamaeleo minor Gunther 1879
  Distribution: Madagascar

Chamaeleo mlanjensis Broadley 1965
  Distribution: Mlanje Mountain, Malawi

Chamaeleo monoceras Boettger 1913
  Distribution: Madagascar

Chamaeleo montium montium Buchholz 1874
  Distribution: Cameroon

Chamaeleo montium feae Boulenger 1906
  Distribution: Fernando Po, Gulf of Guinea

Chamaeleo namaquensis Smith 1831
  Distribution: Western Cape Province north to Angola
  Reference: Burrage (1973); Fitzsimons (1943); Robinson (1978) (1979)

Chamaeleo nasutus Dumeril and Bibron 1836
  Distribution: Madagascar

Chamaeleo oshaughnessyi Gunther 1881
    Distribution: Madagascar

Chamaeleo oustaleti Mocquard 1894
    Distribution: Madagascar; introduced into the Ngong Forest near
            Nairobi, Kenya
    Reference: Bourgat and Brygoo (1968)

Chamaeleo oweni Gray 1831
    Distribution: Nigeria south to Zaire
    Reference: Rand (1958)

Chamaeleo pardalis Cuvier 1829
    Distribution: Madagascar and the Mascarenes
    Reference: Bourgat (1970); Brygoo and Domergue (1969)
    Chamaeleo guentheri Boulenger: Angel (1942)

Chamaeleo parsonii Cuvier 1824
    Distribution: Madagascar and the Mascarenes

Chamaeleo pfefferi Tornier 1900
    Distribution: Cameroon

Chamaeleo polleni Peters 1873
    Distribution: Comoro Islands

Chamaeleo quadricornis Tornier 1899
    Distribution: Mount Manengouba, Cameroon

Chamaeleo quilensis Bocage 1866
    Distribution: Togo east to Uganda, Kenya and Tanzania, south to
            Namibia, Botswana, Transvaal, Swaziland, Natal and
            Zululand
    Reference: Laurent (1964b); Wever (1968)

Chamaeleo rhinoceratus rhinoceratus Gray 1845
    Distribution: Madagascar

Chamaeleo rhinoceratus labordi Grandidier 1872
    Distribution: Madagascar

Chamaeleo rhinoceratus voeltzkowi Boettger 1893
    Distribution: Madagascar

Chamaeleo rudis rudis Boulenger 1906
  Distribution: Uganda, Rwanda and adjacent Zaire
  Reference: Rand (1963)

Chamaeleo rudis schoutedeni Laurent 1952
  Distribution: Kabobo Mountains, Zaire
  Reference: Rand (1963)

Chamaeleo rudis sternfeldi Rand 1963
  Distribution: Tanzania

Chamaeleo schubotzi Sternfeld 1912
  Distribution: Mount Kenya, Kenya
  Reference: Rand (1963)

Chamaeleo senegalensis Daudin 1802
  Distribution: Sudan and Ethiopia south through Uganda, Kenya,
              Tanzania, and Zambia, west through Angola and Zaire
              to Senegal
  Reference: Wever (1968)

Chamaeleo spinosus Matschie 1892
  Distribution: Usambara Mountains, Tanzania

Chamaeleo tempeli Tornier 1899
  Distribution: Uzungwe, Ubena and Ukinga Mountains, Tanzania

Chamaeleo tenuis Matschie 1892
  Distribution: Usambara Mountains, Tanzania

Chamaeleo tigris Kuhl 1820
  Distribution: Seychelles

Chamaeleo verrucosus verrucosus Cuvier 1829
  Distribution: Madagascar

Chamaeleo verrucosus semicristatus Boettger 1894
  Distribution: Madagascar

Chamaeleo werneri Tornier 1899
  Distribution: Uluguru and Uzungwe Mountains, Tanzania
  Reference: Rand (1958)

Chamaeleo wiedersheimi Nieden 1910
    Distribution: Cameroon

Chamaeleo willsii Gunther 1890
    Distribution: Madagascar

Chamaeleo xenorhinus Boulenger 1901
    Distribution: Ruwenzori Mountains, Uganda

Genus: RHAMPHOLEON Gunther 1874
Species typica: *spectrum* Buchholz
Reference: Loveridge (1957)

Rhampholeon brachyurus brachyurus Gunther 1893
    Distribution: Malawi

Rhampholeon brachyurus ionidesi Loveridge 1951
    Distribution: Southeastern Tanzania

Rhampholeon brevicaudatus (Matschie 1892)   *Chamaeleon*
    Distribution: Coastal Tanzania

Rhampholeon kerstenii kerstenii (Peters 1868)   *Chamaeleo*
    Distribution: Kenya and Tanzania

Rhampholeon kerstenii robecchii Boulenger 1892
    Distribution: Somalia and northern Kenya
    Reference: Parker (1942)

Rhampholeon marshalli marshalli Boulenger 1906
    Distribution: Eastern Zimbabwe
    Reference: Broadley (1962b); Fitzsimons (1943)

Rhampholeon marshalli gorongosae Broadley 1971
    Distribution: Mozambique

Rhampholeon nchisiensis (Loveridge 1953)   *Brookesia*
    Distribution: Southwest Tanzania and Malawi

Rhampholeon platyceps platyceps Gunther 1893
    Distribution: Malawi
    Reference: Frank (1951)

Rhampholeon platyceps carri (Loveridge 1953)   *Brookesia*
    Distribution: Mlanje Mountain, Malawi

Rhampholeon spectrum spectrum (Buchholz 1874)   *Chamaeleo*
    Distribution: Cameroon south to Congo; Fernando Po

Rhampholeon spectrum boulengeri Steindachner 1911
    Distribution: Northern and eastern Zaire, Rwanda, Burundi, Uganda,
        Kenya and Tanzania

Rhampholeon temporalis (Matschie 1892)   *Chamaeleo*
    Distribution: Usambara Mountains, Tanzania

Chapter 8

FAMILY SCINCIDAE

Order: Sauria
Suborder: Autarchoglossa
Infraorder: Scincomorpha
Superfamily: Scincoidea
Family: Scincidae
Subfamily: Feylininae

Genus: CHABANAUDIA Witte and Laurent 1943
Species typica: *boulengeri* Chabanaud

Chabanaudia boulengeri (Chabanaud 1917)   *Feylinia*
    Distribution: Gabon

Genus: FEYLINIA Gray 1845
Species typica: *currori* Gray
Reference: Witte and Laurent (1943b)

Feylinia currori Gray 1845
    Distribution: Cameroon, Gabon, Congo, Zaire and Angola
    Reference: Gasc (1965); Lanza and Vanni (1976); Laurent (1964b)

Feylinia elegans (Hallowell 1852)   *Acontias*
    Distribution: Gabon, Congo, Zaire, Angola and Uganda
    Reference:
    Feylinia elegans grandisquamis Muller: Laurent (1964b); Witte and
                         Laurent (1943b)

Feylinia polylepis Bocage 1887
   Distribution: Principe Island, Gulf of Guinea

Subfamily: Acontinae

Genus: ACONTIAS Cuvier 1817
Species typica: *meleagris* Linnaeus
Reference: Broadley and Greer (1969); Fitzsimons (1943); Witte and Laurent
      (1943b)

Acontias breviceps Essex 1925
   Distribution: Eastern Cape Province and eastern Transvaal

Acontias gracilicauda gracilicauda Essex 1925
   Distribution: Cape Province, Orange Free State and Transvaal

Acontias gracilicauda namaquensis Hewitt 1938
   Distribution: Little Namaqualand

Acontias lineatus lineatus Peters 1879
   Distribution: Western Cape Province and southern Namibia
   Reference:
   Acontias lineatus orangensis Hewitt: Fitzsimons (1943); Witte and
                        Laurent (1943b)

Acontias lineatus grayi Boulenger 1887
   Distribution: Southwestern Cape Province

Acontias lineatus tristis Werner 1911
   Distribution: Little Namaqualand

Acontias litoralis Broadley and Greer 1969
   Distribution: Western coast of South Africa

Acontias meleagris meleagris (Linnaeus 1758)   *Anguis*
   Distribution: Southwestern Cape Province

Acontias meleagris orientalis Hewitt 1938
   Distribution: Eastern Cape Province
   Reference:
   Acontias meleagris lineicauda Hewitt: Fitzsimons (1943); Witte and
                        Laurent (1943b)

Acontias percivali percivali Loveridge 1935
    Distribution: Vicinity of Voi, Kenya

Acontias percivali occidentalis Fitzsimons 1941
    Distribution: Southern Angola, Namibia, Botswana, Zimbabwe and
        northern Transvaal

Acontias percivali tasmani Hewitt 1937
    Distribution: Eastern Cape Province

Acontias plumbeus Bianconi 1849
    Distribution: Mozambique, northern Natal, eastern Zimbabwe,
        Transvaal and Swaziland

Genus: ACONTOPHIOPS Sternfeld 1912
Species typica: *lineatus* Sternfeld

Acontophiops lineatus Sternfeld 1912
    Distribution: Northern Transvaal
    Reference: Fitzsimons (1943); Witte and Laurent (1943b)

Genus: TYPHLOSAURUS Wiegmann 1834
Species typica: *caecus* Cuvier
Reference: Broadley (1968b); Cloudsley-Thompson and Chadwick (1964);
        Fitzsimons (1943); Witte and Laurent (1943b)

Typhlosaurus aurantiacus aurantiacus (Peters 1854)   *Typhline*
    Distribution: Southern Mozambique, northern Zululand and
        southeast Zimbabwe
    Reference: Van den Heever (1976)

Typhlosaurus aurantiacus fitzsimonsi Broadley 1968
    Distribution: Northeastern Transvaal

Typhlosaurus braini Haacke 1964
    Distribution: Namibia

Typhlosaurus caecus (Cuvier 1817)   *Acontias*
    Distribution: Coastal western Cape Province

Typhlosaurus cregoi cregoi Boulenger 1903
    Distribution: Northern Transvaal

Typhlosaurus cregoi bicolor Hewitt 1929
Distribution: East and southeast Zimbabwe

Typhlosaurus gariepensis Fitzsimons 1941
Distribution: Northern Cape Province
Reference: Kim, Gorman and Huey (1978)

Typhlosaurus lineatus lineatus Boulenger 1887
Distribution: Western Namibia, northern Cape Province and Botswana
Reference: Brain (1959a); Kim, Gorman and Huey (1978)

Typhlosaurus lineatus jappi Broadley 1968
Distribution: Barotseland, Zambia and adjoining eastern Angola

Typhlosaurus lineatus subtaeniatus Broadley 1968
Distribution: Great Saltpan, northern Transvaal

Typhlosaurus meyeri Boettger 1894
Distribution: Coastal Namibia
Reference:
Typhlosaurus plowesi Fitzsimons: Fitzsimons (1943)

Typhlosaurus vermis Boulenger 1887
Distribution: Western Cape Province

Subfamily: Scincinae

Genus: AMPHIGLOSSUS Dumeril and Bibron 1839
Species typica: *astrolabi* Dumeril and Bibron
Reference: Angel (1942)

Amphiglossus andranovahensis (Angel 1933)   *Scelotes*
Distribution: Madagascar

Amphiglossus ankodabensis (Angel 1930)   *Scelotes*
Distribution: Madagascar

Amphiglossus ardouini (Mocquard 1897)   *Sepsina*
Distribution: Madagascar

Amphiglossus astrolabi astrolabi Dumeril and Bibron 1839
    Distribution: Madagascar

Amphiglossus astrolabi boettgeri (Angel 1942)   *Scelotes*
    Distribution: Madagascar

Amphiglossus crenni (Mocquard 1906)   *Sepsina*
    Distribution: Madagascar

Amphiglossus decaryi (Angel 1930)   *Scelotes*
    Distribution: Madagascar

Amphiglossus elongatus (Angel 1933)   *Scelotes*
    Distribution: Madagascar

Amphiglossus frontoparietalis (Boulenger 1889)   *Sepsina*
    Distribution: Madagascar

Amphiglossus gastrostictus (O'Shaughnessy 1879)   *Gongylus*
    Distribution: Madagascar

Amphiglossus igneocaudatus (Grandidier 1867)   *Gongylus*
    Distribution: Madagascar

Amphiglossus intermedius (Boettger 1913)   *Scelotes*
    Distribution: Madagascar

Amphiglossus johannae (Gunther 1880)   *Gongylus*
    Distribution: Comoro Islands

Amphiglossus macrocercus (Gunther 1882)   *Gongylus*
    Distribution: Madagascar

Amphiglossus macrolepis (Boulenger 1888)   *Scelotes*
    Distribution: Madagascar

Amphiglossus melanopleura (Gunther 1877)   *Gongylus*
    Distribution: Madagascar

Amphiglossus melanurus (Gunther 1877)   *Gongylus*
    Distribution: Madagascar

Amphiglossus mouroundavae (Grandidier 1872)   *Gongylus*
    Distribution: Madagascar

Amphiglossus ornaticeps (Boulenger 1896)   *Sepsina*
    Distribution: Madagascar

Amphiglossus poecilopus (Barbour and Loveridge 1928)   *Scelotes*
    Distribution: Madagascar

Amphiglossus polleni (Grandidier 1869)   *Gongylus*
    Distribution: Madagascar

Amphiglossus praeornatus (Angel 1938)   *Scelotes*
    Distribution: Madagascar

Amphiglossus reticulatus (Kaudern 1922)   *Sepsina*
    Distribution: Madagascar

Amphiglossus splendidus (Grandidier 1872)   *Gongylus*
    Distribution: Madagascar

Amphiglossus trilineatus (Angel 1949)   *Gongylus*
    Distribution: Madagascar

Amphiglossus vulsini (Barbour 1918)   *Sepsina*
    Distribution: Madagascar

Amphiglossus waterloti (Angel 1930)   *Scelotes*
    Distribution: Madagascar

Genus: CHALCIDES Laurenti 1768
Species typica: *chalcides* Linnaeus
Reference: Pasteur (1981); Pasteur and Bons (1960)

Chalcides armitagei E. G. Boulenger 1921
    Distribution: Gambia

Chalcides chalcides (Linnaeus 1758)   *Lacerta*
    Distribution: Northern Morocco, Algeria, Tunisia, and northwest
            Libya
    Reference: Arnold and Burton (1978); Schmidt and Inger (1957)

Chalcides colosii Lanza 1957
    Distribution: Northern Morocco

Chalcides ebneri Werner 1931
   Distribution: vicinity of Fez, Morocco

Chalcides mauritanicus (Dumeril and Bibron 1839)   *Heteromeles*
   Distribution: Coastal Algeria

Chalcides mionecton (Boettger 1873)   *Seps*
   Distribution: Western lowlands of Morocco

Chalcides ocellatus (Forskal 1775)   *Lacerta*
   Distribution: Kenya and Somalia north to Egypt then west to Morocco
          to include Chad and Niger
   Reference: Arnold and Burton (1978); Badir (1959) (1968a) (1968b);
          Badir and Hussein (1965); Khalil (1951); Parker (1942);
          Richter (1933); Tercafs (1962); Trost (1953); Vogel (1964)

Chalcides polylepis polylepis Boulenger 1890
   Distribution: Morocco

Chalcides polylepis occidentalis Steindachner 1900
   Distribution: Eastern Canary Islands

Chalcides pulchellus Mocquard 1906
   Distribution: Southwestern Upper Volta

Chalcides ragazzii ragazzii Boulenger 1890
   Distribution: Eritrea, Ethiopia

Chalcides ragazzii bottegi Boulenger 1898
   Distribution: Ethiopia, Somalia, Kenya, Sudan and Chad

Chalcides sexlineatus Steindachner 1891
   Distribution: Gran Canary and Madeira

Chalcides thierryi Tornier 1901
   Distribution: Northern Ghana east to central Nigeria

Chalcides viridanus (Gravenhorst 1851)   *Gongylus*
   Distribution: Western Canary Islands (excluding Gran Canary) and
          Madeira

Genus: CRYPTOPOSCINCUS Mocquard 1906
Species typica: *minimus* Mocquard

Cryptoposcincus minimus Mocquard 1906
    Distribution: Madagascar
    Reference: Angel (1942)

Genus: EUMECES Wiegmann 1834
Species typica: *pavimentatus* Geoffroy

Eumeces algeriensis algeriensis Peters 1864
    Distribution: Northern Morocco and coastal Algeria
    Reference: Pasteur and Bons (1960); Vogel (1964)

Eumeces algeriensis meridionalis Doumergue 1900
    Distribution: Atlas Mountains of Morocco and Algeria
    Reference: Pasteur and Bons (1960)

Eumeces schneideri (Daudin 1802)   *Scincus*
    Distribution: Egypt
    Reference: Eiselt (1940); Marx (1968); Vogel (1964)

Genus: GONGYLOMORPHUS Fitzinger 1843
Species typica: *bojeri* Desjardin

Gongylomorphus bojeri (Desjardin 1831)   *Scincus*
    Distribution: Mascarenes

Genus: GRANDIDIERINA Mocquard 1894
Species typica: *rubrocaudata* Grandidier
Reference: Angel (1942)

Grandidierina fierinensis (Grandidier 1869)   *Scelotes*
    Distribution: Madagascar

Grandidierina lineata Mocquard 1901
    Distribution: Madagascar

Grandidierina petiti Angel 1924
    Distribution: Madagascar

Grandidierina rubrocaudata (Grandidier 1869)   *Acontias*
    Distribution: Madagascar

Genus: JANETAESCINCUS Greer 1970
Species typica: *braueri* Boettger

Janetaescincus braueri (Boettger 1896)   *Scelotes*
    Distribution: Seychelles

Janetaescincus veseyfitzgeraldi (Parker 1948)   *Scelotes*
    Distribution: Seychelles

Genus: MALACONTIAS Greer 1970
Species typica: *holomelas* Gunther
Reference: Angel (1942)

Malacontias hildebrandtii (Peters 1880)   *Acontias*
    Distribution: Madagascar

Malacontias holomelas (Gunther 1877)   *Acontias*
    Distribution: Madagascar

Genus: MELANOSEPS Boulenger 1887
Species typica: *atra* Gunther
Reference: Loveridge (1957); Witte and Laurent (1943b)

Melanoseps ater ater (Gunther 1873)   *Herpetosaura*
    Distribution: Mozambique, southern Malawi and Zambia
    Reference: Broadley (1971e); Loveridge (1953)

Melanoseps ater longicauda Tornier 1900
    Distribution: Southern Kenya and northern Tanzania

Melanoseps ater matengoensis Loveridge 1942
    Distribution: Northwest Mozambique and southwest Tanzania

Melanoseps ater misukuensis Loveridge 1953
    Distribution: Misuku Mountains, Malawi

Melanoseps ater rondoensis Loveridge 1942
    Distribution: Rondo Plateau, Tanzania

Melanoseps ater uzungwensis Loveridge 1942
    Distribution: Uzungwe Mountains, Tanzania

Melanoseps occidentalis (Peters 1877)   *Herpetosaura*
    Distribution: Cameroon south to Angola
    Reference: Laurent (1964b)

Genus: PAMELAESCINCUS Greer 1970
Species typica: *gardinieri* Boulenger

Pamelaescincus gardinieri (Boulenger 1909)   *Scelotes*
    Distribution: Seychelles

Genus: PARACONTIAS Mocquard 1894
Species typica: *brocchii* Mocquard
Reference: Angel (1942)

Paracontias brocchii Mocquard 1894
    Distribution: Madagascar

Paracontias milloti Angel 1949
    Distribution: Madagascar

Paracontias rothschildi Mocquard 1905
    Distribution: Madagascar

Genus: PROSCELOTES Witte and Laurent 1943
Species typica: *eggeli* Tornier
Reference: Witte and Laurent (1943b)

Proscelotes aenea (Barbour and Loveridge 1928)   *Scelotes*
    Distribution: Mozambique

Proscelotes arnoldi (Hewitt 1932)   *Sepsina*
    Distribution: Malawi and Zimbabwe
    Reference: Fitzsimons (1943); Loveridge (1953)

Proscelotes eggeli (Tornier 1902)   *Scelotes*
    Distribution: Tanzania
    Reference: Loveridge (1957)

Genus: PSEUDACONTIAS Bocage 1889
Species typica: *madagascariensis* Bocage

Pseudacontias madagascariensis Bocage 1889
    Distribution: Madagascar
    Reference: Angel (1942)

Genus: PYGOMELES Grandidier 1867
Species typica: *braconnieri* Grandidier
Reference: Angel (1942)

Pygomeles braconnieri Grandidier 1867
    Distribution: Madagascar

Pygomeles petteri Pasteur and Paulin 1962
    Distribution: Madagascar

Pygomeles trivittatus Boulenger 1896
    Distribution: Madagascar

Genus: SCELOTES Fitzinger 1826
Species typica: *bipes* Linnaeus
Reference: Fitzsimons (1943); Witte and Laurent (1943b)

Scelotes anguina (Boulenger 1887)   *Herpetoseps*
    Distribution: Coastal eastern Cape Province

Scelotes arenicolor (Peters 1854)   *Herpetosaura*
    Distribution: Zululand, Swaziland and southern Mozambique

Scelotes bicolor (Smith 1849)   *Lithophilus*
    Distribution: Little Namaqualand

Scelotes bidigittata Fitzsimons 1930
    Distribution: Eastern Transvaal

Scelotes bipes bipes (Linnaeus 1758)   *Anguis*
    Distribution: Southwestern Cape Province
    Reference: Witte and Laurent (1943b, part)

Scelotes bipes sexlineatus (Harlan 1824)   *Seps*
    Distribution: Northwestern Cape Province
    Reference:
    Scelotes bipes bipes: Witte and Laurent (1943b, part)
    Scelotes bipes namaquensis Fitzsimons: Witte and Laurent (1943b)

Scelotes brevipes Hewitt 1925
   Distribution: Zululand, eastern Transvaal, Swaziland and southern
            Mozambique

Scelotes caffer (Peters 1861)   *Sepomorphus*
   Distribution: Cape Province
   Reference:
   Herpetosaura tridactyla (Boulenger): Witte and Laurent (1943b)

Scelotes capensis (Smith 1849)   *Gongylus*
   Distribution: Western coast of South Africa
   Reference:
   Herpetosaura capensis depressa (Hewitt): Witte and Laurent (1943b)
   Herpetosaura weberi (Roux): Witte and Laurent (1943b)

Scelotes gronovi (Daudin 1802)   *Seps*
   Distribution: Coastal southwestern Cape Province

Scelotes guentheri Boulenger 1887
   Distribution: Durban

Scelotes inornata inornata (Smith 1849)   *Lithophilus*
   Distribution: Coastal Natal
   Reference: Witte and Laurent (1943b, part)
   Herpetosaura natalensis (Hewitt): Witte and Laurent (1943b)

Scelotes inornata mossambica (Peters 1882)   *Herpetosaura*
   Distribution: Northern Zululand and southern Mozambique
   Reference:
   Herpetosaura inornata: Witte and Laurent (1943b, part)

Scelotes kasneri Fitzsimons 1939
   Distribution: Lamberts Bay, Cape Province

Scelotes limpopoensis Fitzsimons 1930
   Distribution: Northern Transvaal and Zimbabwe

Scelotes mira (Roux 1907)   *Herpetosaura*
   Distribution: Eastern Transvaal and Swaziland

Scelotes poensis Bocage 1895
   Distribution: Fernando Po, Gulf of Guinea

Scelotes schebeni Sternfeld 1910
   Distribution: Namibia

Scelotes uluguruensis Barbour and Loveridge 1928
  Distribution: Tanzania
  Reference: Loveridge (1957)

Genus: SCINCOPUS Peters 1864
Species typica: *fasciatus* Peters

Scincopus fasciatus (Peters 1864)   *Scincus*
  Distribution: Saharan North Africa from Mauritania to Sudan

Genus: SCINCUS Laurenti 1768
Species typica: *scincus* Linnaeus
Reference: Arnold and Leviton (1977)

Scincus scincus scincus (Linnaeus 1758)   *Lacerta*
  Distribution: Egypt
  Reference: Badir (1959); Badir and Hussein (1965); El-Toubi (1938);
       Gabe and Saint Girons (1965); Richter (1933)

Scincus scincus albifasciatus Boulenger 1890
  Distribution: Western Sahara, Mauritania, Senegal, Mali, Niger and
          southeast Libya

Scincus scincus cucullatus Werner 1914
  Distribution: Northwest Libya, Tunisia and northwest Algeria

Scincus scincus laterimaculatus Werner 1914
  Distribution: Northwest Algeria and adjacent Morocco
  Reference: Pasteur and Bons (1960)

Genus: SCOLECOSEPS Loveridge 1920
Species typica: *boulengeri* Loveridge
Reference: Witte and Laurent (1943b)

Scolecoseps acontias (Werner 1914)   *Melanoseps*
  Distribution: Tanzania

Scolecoseps boulengeri Loveridge 1920
  Distribution: Mozambique

Genus: SEPSINA Bocage 1866
Species typica: *angolensis* Bocage
Reference: Witte and Laurent (1943b)

Sepsina alberti Hewitt 1929
Distribution: Namibia
Reference: Fitzsimons (1943)

Sepsina angolensis Bocage 1866
Distribution: Namibia and Zambia north to southern Zaire
Reference: Broadley (1971e); Fitzsimons (1943); Laurent (1964b)
Sepsina grammica Cope: Witte and Laurent (1943b)

Sepsina bayoni (Bocage 1866)   *Dumerilia*
Distribution: Angola and western Zaire

Sepsina copei Bocage 1873
Distribution: Angola

Sepsina tetradactyla tetradactyla Peters 1874
Distribution: Tanzania south to Malawi and Zambia
Reference: Broadley (1971e); Loveridge (1953) (1957)

Sepsina tetradactyla hemptinnei Witte 1933
Distribution: Western Tanzania and adjacent Zaire
Reference: Loveridge (1957)

Genus: SPHENOPS Wagler 1830
Species typica: *sepsoides* Audouin
Reference: Pasteur and Bons (1960)

Sphenops boulengeri (Anderson 1892)   *Chalcides*
Distribution: Saharan Morocco, Algeria and Tunisia

Sphenops delislei (Lataste 1880)   *Scincus*
Distribution: Central Sahara

Sphenops sepsoides (Audouin 1827)   *Scincus*
Distribution: Egypt
Reference: Khalil and Hussein (1962 (1963); Marx (1968)

Sphenops sphenopsiformis (Dumeril 1856)   *Chalcides*
Distribution: Southern Morocco south to Senegal and Mauritania

Genus: TYPHLACONTIAS Bocage 1873
Species typica: *punctatissimus* Bocage
Reference: Witte and Laurent (1943b)

Typhlacontias bogerti Laurent 1964
    Distribution: Southern Angola and Namibia
    Reference: Haacke (1965)

Typhlacontias brevipes Fitzsimons 1938
    Distribution: Central Namib Desert
    Reference: Haacke (1964)

Typhlacontias gracilis Roux 1907
    Distribution: Zimbabwe and Zambia

Typhlacontias ngamiensis Fitzsimons 1932
    Distribution: Zimbabwe and Botswana
    Reference: Fitzsimons (1943); Haacke (1965)

Typhlacontias punctatissimus Bocage 1873
    Distribution: Southern Angola

Typhlacontias rohani Angel 1924
    Distribution: Southeastern Angola

Genus: VOELTZKOWIA Boettger 1893
Species typica: *mira* Boettger

Voeltzkowia mira Boettger 1893
    Distribution: Madagascar

Subfamily: Lygosominae

Genus: COPHOSCINCOPUS Mertens 1934
Species typica: *dura* Cope

Cophoscincopus durus (Cope 1862)   *Tiliqua*
    Distribution: West Africa
    Reference: Mudrack (1977)

Genus: CRYPTOBLEPHARUS Wiegmann 1834
Species typica: *boutonii* Desjardin
Reference: Fuhn (1970); Mertens (1931)

Cryptoblepharus africanus (Sternfeld 1918)   *Ablepharus*
    Distribution: Coastal Somalia, Kenya and Tanzania
    Reference: Lanza and Carfi (1968); Loveridge (1957)

Cryptoblepharus aldabrae (Sternfeld 1918)   *Ablepharus*
    Distribution: Aldabra

Cryptoblepharus ater (Boettger 1913)   *Ablepharus*
    Distribution: Comoro Islands

Cryptoblepharus bitaeniatus (Boettger 1913)   *Ablepharus*
    Distribution: Europa Island

Cryptoblepharus boutonii (Dsjardin 1831)   *Ablepharus*
    Distribution: Mascarenes

Cryptobelpharus caudatus (Sternfeld 1918)   *Ablepharus*
    Distribution: Juan de Nova Island

Cryptoblepharus cognatus (Boettger 1881)   *Ablepharus*
    Distribution: Madagascar

Cryptoblepharus degrijsi (Mertens 1928)   *Ablepharus*
    Distribution: Comoro Islands

Cryptoblepharus gloriosus (Stejneger 1893)   *Ablepharus*
    Distribution: Madagascar

Cryptoblepharus mayottensis (Mertens 1928)   *Ablepharus*
    Distribution: Comoro Islands

Cryptoblepharus mohelicus (Mertens 1928)   *Ablepharus*
    Distribution: Comoro Islands

Cryptoblepharus voeltzkowi (Sternfeld 1918)   *Ablepharus*
    Distribution: Madagascar

Genus: EUMECIA Bocage 1870
Species typica: *anchietae* Bocage
Reference: Greer (1967)

Eumecia anchietae anchietae Bocage 1870
Distribution: Angola, Zaire, Kenya, Tanzania and Zambia
Reference: Broadley (1971e); Loveridge (1957)

Eumecia anchietae major Laurent 1964
Distribution: Angola

Eumecia anchietae wittei Laurent 1964
Distribution: Southern Zaire

Eumecia johnstoni (Boulenger 1897)   *Lygosoma*
Distribution: Malawi

Genus: LEIOLOPISMA Dumeril and Bibron 1839
Species typica: *telfairii* Desjardin

Leiolopisma telfairii (Desjardin 1831)   *Scincus*
Distribution: Round Island

Genus: LYGOSOMA Hardwick and Gray 1827
Species typica: *quadrupes* Linnaeus
Reference: Greer (1977)

Lygosoma afer (Peters 1854)   *Eumeces*
Distribution: Sudan, Ethiopia, Somalia, Uganda, Kenya, Tanzania,
            eastern Zambia, Malawi, Mozambique and Zaire
Reference: Broadley (1966a); Lanza and Carfi (1968); Lanza and Vanni
          (1976)

Lygosoma fernandi (Burton 1836)   *Tiliqua*
Distribution: Guinea east to Zaire, Angola and Uganda
Reference: Laurent (1964b); Loveridge (1957)

Lygosoma grandisonianum (Lanza and Carfi 1966)   *Mochlus*
Distribution: Somalia
Reference: Lanza and Carfi (1968)

Lygosoma guineense (Peters 1879)   *Euprepes*
Distribution: Guinea east to northern Zaire and Uganda

Lygosoma laeviceps laeviceps (Peters 1874)   *Euprepes*
Distribution: Somalia
Reference: Parker (1942)

Lygosoma laeviceps paedocarinatum (Lanza and Carfi 1968)   *Mochlus*
    Distribution: Somalia and Ethiopia

Lygosoma mabuiiforme (Loveridge 1935)   *Riopa*
    Distribution: Kenya and Somalia
    Reference: Lanza and Carfi (1968)

Lygosoma pembanum Boettger 1913
    Distribution: Pemba Island

Lygosoma productum Boulenger 1909
    Distribution: Somalia
    Reference: Lanza and Carfi (1968)

Lygosoma simonettai Lanza 1979
    Distribution: Somalia

Lygosoma sundevalli sundevalli (Smith 1849)   *Eumeces*
    Distribution: Angola, Namibia, Botswana, Transvaal, Mozambique,
        Zimbabwe and Zambia
    Reference: Broadley (1966a); Haacke (1965); Lanza and Carfi (1968);
        Laurent (1964b)

Lygosoma sundevalli modestum (Gunther 1880)   *Sepacontias*
    Distribution: Malawi, Tanzania and Kenya

Lygosoma sundevalli somalicum (Parker 1942)   *Riopa*
    Distribution: Somalia

Lygosoma tanae (Loveridge 1935)   *Riopa*
    Distribution: Coastal Kenya and Somalia
    Reference: Lanza and Carfi (1968)

Lygosoma vinciguerrae Parker 1932
    Distribution: Somalia
    Reference: Lanza and Carfi (1968)

Genus: MABUYA Fitzinger 1826
Species typica: *mabouya* Lacepede

Mabuya acutilabris (Peters 1862)   *Euprepes*
    Distribution: Northwest Cape Province north to Zaire
    Reference: Fitzsimons (1943); Laurent (1964b)

Mabuya affinis (Gray 1839)   *Tiliqua*
    Distribution: Angola north along African coast to Senegal
    Reference: Hoogmoed (1974)
    Mabuya blandingi (Hallowell): Barbault (1974a); Hoogmoed (1973a)
            (1973b)

Mabuya albilabris (Hallowell 1857)   *Euprepes*
    Distribution: Guinea east along the coast to Gabon
    Reference: Hoogmoed (1973a) (1973b) (1974)

Mabuya angolensis Monard 1937
    Distribution: Angola
    Reference: Laurent (1964b)

Mabuya aureopunctata (Grandidier 1867)   *Euprepes*
    Distribution: Madagascar
    Reference: Angel (1942)

Mabuya bayonii bayonii (Bocage 1872)   *Euprepes*
    Distribution: Zaire and Angola
    Reference: Laurent (1964b)

Mabuya bayonii huilensis Laurent 1964
    Distribution: Angola

Mabuya bayonii keniensis Loveridge 1956
    Distribution: Kenya and Tanzania

Mabuya bensonii (Peters 1867)   *Euprepes*
    Distribution: Liberia and Sierra Leone

Mabuya betsileana Mocquard 1906
    Distribution: Madagascar
    Reference: Angel (1942)

Mabuya binotata (Bocage 1867)   *Euprepes*
    Distribution: Northern Namibia and southern Angola
    Reference: Fitzsimons (1943); Laurent (1964b)

Mabuya bocagii bocagii Boulenger 1887
    Distribution: Angola, Zambia and Zimbabwe

Mabuya bocagii brauni Tornier 1902
    Distribution: Tanzania

Mabuya bocagii mlanjensis Loveridge 1953
    Distribution: Malawi

Mabuya boettgeri Boulenger 1887
    Distribution: Madagascar
    Reference: Angel (1942)

Mabuya boulengeri Sternfeld 1911
    Distribution: Southeastern Tanzania south through Mozambique to the
            Save River west into southeast Malawi and eastern
            Zimbabwe
    Reference: Broadley (1974); Loveridge (1957)

Mabuya brevicollis (Wiegmann 1837)   *Euprepes*
    Distribution: Sudan, Ethiopia, Somalia, Kenya, Uganda and Tanzania
    Reference: Lanza and Carfi (1968); Loveridge (1957); Matz (1973);
            Parker (1942)

Mabuya buettneri Matschie 1893
    Distribution: Ivory Coast, Ghana, Togo and northern Zaire
    Reference: Barbault (1967) (1971) (1974d) (1975) (1976); Hoogmoed
            (1973a) (1973b) (1974)

Mabuya capensis (Gray 1831)   *Tiliqua*
    Distribution: South Africa
    Reference: Fitzsimons (1943)

Mabuya chimbana Boulenger 1887
    Distribution: Southwestern Angola
    Reference: Broadley (1975b); Laurent (1964b)

Mabuya delalandii (Dumeril and Bibron 1839)   *Euprepes*
    Distribution: Cape Verde Islands

Mabuya elegans (Peters 1854)   *Euprepes*
    Distribution: Madagascar
    Reference: Angel (1942)

Mabuya ferrarai Lanza 1978
    Distribution: Coastal southern Somalia

Mabuya fogoensis (O'Shaughnessy 1874)   *Euprepes*
    Distribution: Cape Verde Islands

Mabuya gravenhorstii (Dumeril and Bibron 1839)   *Euprepes*
　　Distribution: Madagascar
　　Reference: Angel (1942)

Mabuya hemmingi Laurent and Gans 1965
　　Distribution: Somalia
　　Reference: (Lanza and Carfi 1968)

Mabuya hildae Loveridge 1953
　　Distribution: Malawi

Mabuya hildebrandtii (Peters 1874)   *Euprepes*
　　Distribution: Somalia
　　Reference: Lanza and Carfi (1968)

Mabuya hoeschi Mertens 1954
　　Distribution: Angola

Mabuya homalocephala homalocephala (Wiegmann 1828)   *Scincus*
　　Distribution: Cape Province
　　Reference: Fitzsimons (1943)

Mabuya homalocephala depressa (Peters 1854)   *Euprepes*
　　Distribution: Natal and Zululand north to Zambesi River, Mozambique
　　Reference: Fitzsimons (1943)

Mabuya homalocephala peringueyi Boulenger 1888
　　Distribution: Northwestern Cape Province and Namibia
　　Reference: Fitzsimons (1943)

Mabuya homalocephala smithi (Gray 1845)   *Euprepis*
　　Distribution: Eastern Cape Province
　　Reference: Fitzsimons (1943)

Mabuya irregularis Lonnberg 1922
　　Distribution: Uganda and Kenya
　　Reference: Andren (1976); Loveridge (1957)

Mabuya ivensii ivensii Bocage 1879
　　Distribution: Angola
　　Reference: Greer (1977); Horton (1972)

Mabuya ivensii septemlineata Laurent 1964
　　Distribution: Angola

Mabuya lacertiformis (Peters 1854)   *Euprepes*
    Distribution: Southern Lake Malawi south through Tete, Mozambique
            to eastern Zimbabwe, west along the south bank of the
            Zambesi to Wankie District; southwestern Angola
    Reference: Broadley (1971e) (1975b)
    Mabuya damarana (Peters): Fitzsimons (1943, part)
    Mabuya damarana rhodesiana Broadley: Broadley (1960c)

Mabuya laevis Boulenger 1907
    Distribution: Northern Namibia and southwestern Angola
    Reference: Greer (1977); Laurent (1964b); Steyn and Mitchell (1965)

Mabuya maculilabris maculilabris (Gray 1845)   *Euprepes*
    Distribution: Liberia east to Uganda, Somalia, Kenya and south to
            Angola, Zambia, Malawi and Mozambique
    Reference: Barbault (1974b) (1975) (1976); Broadley (1971e) (1974);
            Hoogmoed (1973a) (1973b) (1974); Lanza and Carfi (1968);
            Lanza and Vanni (1976); Laurent (1964b)

Mabuya maculilabris albotaeniata Boettger 1913
    Distribution: Pemba Island, Tanzania
    Reference: Broadley (1974); Loveridge (1957)

Mabuya maculilabris casuarinae Broadley 1974
    Distribution: Casuarina Island, Mozambique

Mabuya maculilabris comorensis (Peters 1854)   *Euprepes*
    Distribution: Comoro Islands
    Reference: Angel (1942); Broadley (1974)

Mabuya maculilabris infralineata Boettger 1913
    Distribution: Europa Island
    Reference: Angel (1942)

Mabuya madagascariensis Mocquard 1908
    Distribution: Madagascar
    Reference: Angel (1942)

Mabuya megalura (Peters 1878)   *Euprepes*
    Distribution: Ethiopia, Somalia, Uganda, Kenya, Tanzania, Rwanda,
            Burundi and eastern Zaire
    Reference: Broadley (1971e); Herlant and Pasteels (1955); Loveridge
            (1957)

Mabuya occidentalis (Peters 1867)   *Euprepes*
    Distribution: Northwest Cape Province north to southern Angola
    Reference: Fitzsimons (1943); Huey and Pianka (1977); Laurent (1964b)

Mabuya perrotetii (Dumeril and Bibron 1839)   *Euprepes*
    Distribution: Senegal east along the coast to Northern Zaire and Uganda
    Reference: Barbault (1975); Hoogmoed (1974)

Mabuya planifrons (Peters 1878)   *Euprepes*
    Distribution: Ethiopia, Somalia, Kenya, Tanzania and Zaire
    Reference: Broadley (1971e); Lanza and Carfi (1968); Loveridge (1957);
            Parker (1942)

Mabuya polytropis Boulenger 1903
    Distribution: Liberia, Ghana, Cameroon, Gabon, Zaire and Fernando Po
    Reference: Hoogmoed (1974)

Mabuya quinquetaeniata quinquetaeniata (Lichtenstein 1823)   *Scincus*
    Distribution: Egypt south to northern Uganda and west to Algerian
            Sahara
    Reference: Cloudsley-Thompson (1965); Lanza and Carfi (1968);
            Parker (1942)

Mabuya quinquetaeniata margaritifer (Peters 1854)   *Euprepes*
    Distribution: Mozambique south of the Zambesi, Malawi, Zimbabwe,
            Zambia, eastern Botswana and northeastern South Africa
    Reference: Broadley (1971); Fitzsimons (1943); Simbotwe (1980)

Mabuya quinquetaeniata obsti Werner 1913
    Distribution: Mozambique north of the Zambesi, Tanzania, Kenya and
            Uganda

Mabuya quinquetaeniata scharica Sternfeld 1917
    Distribution: Senegal east to Congo and Central African Republic
    Reference: Cansdale (1951); Hoogmoed (1973a) (1973d) (1974); Lanza
            and Vanni (1976)

Mabuya rodenburgi Hoogmoed 1974
    Distribution: Ghana and Nigeria
    Reference: Gartshore (1978)

Mabuya sakalava (Grandidier 1872)   *Euprepes*
    Distribution: Madagascar
    Reference: Angel (1942)

Mabuya seychellensis (Dumeril and Bibron 1839)   *Euprepes*
    Distribution: Seychelles
    Reference: Crawford and Thorpe (1979)

Mabuya spilogaster (Peters 1882)   *Euprepes*
  Distribution: Southern Angola and Namibia east to Mozambique
  Reference: Huey and Pianka (1977); Laurent (1964b)

Mabuya stanjeri Gray 1845
  Distribution: Cape Verde Islands

Mabuya striata (Peters 1844)   *Tropidolepisma*
  Distribution: South Africa north to southern Angola, Uganda, Ethiopia
            and Somalia; Comoro Islands
  Reference: Angel (1942); Fitzsimons (1943); Hirji (1977); Huey and
            Pianka (1977); Lanza and Carfi (1968); Robertson and others
            (1965); Simbotwe (1980); Stebbins (1961)

Mabuya sulcata (Peters 1867)   *Euprepes*
  Distribution: Cape Province north to southern Angola
  Reference: Fitzsimons (1943)
  Mabuya sulcata ansorgei Boulenger: Laurent (1964b)

Mabuya vaillantii Boulenger 1887
  Distribution: Cape Verde Islands

Mabuya varia varia (Peters 1867)   *Euprepes*
  Distribution: Southern Sudan and Somalia south to Mozambique,
            west through Zimbabwe, Bostwana, Namibia, Angola
            and Zaire
  Reference: Andren (1976); Fitzsimons (1943); Lanza and Carfi (1968)
  Mabuya damarana (Peters): Fitzsimons (1943, part)

Mabuya varia nyikae Loveridge 1953
  Distribution: Malawi

Mabuya variegata variegata (Peters 1869)   *Euprepes*
  Distribution: Republic of South Africa
  Reference: Broadley (1975b); Huey and Pianka (1977)
  Mabuya damarana (Peters): Fitzsimons (1943, part)

Mabuya variegata punctulata (Bocage 1872)   *Euprepes*
  Distribution: Southwest Angola, Namibia, Zambia, Botswana, northern
            Cape Province, northwest Zimbabwe, Transvaal and
            Mozambique
  Reference: Broadley (1975b); Fitzsimons (1943)
  Mabuya damarana (Peters): Fitzsimons (1943b, part)

Mabuya vittata (Olivier 1804)   *Scincus*
    Distribution: Egypt

Mabuya wrightii Boulenger 1887
    Distribution: Seychelles

Genus: MACROSCINCUS Bocage 1873
Species typica: *coctei* Dumeril and Bibron

Macroscincus coctei (Dumeril and Bibron 1839)   *Euprepis*
    Distribution: Cape Verde Islands
    Reference: Greer (1976)

Genus: PANASPIS Cope 1868
Species typica: *aeneus* Cope = *cabindae* Bocage
Reference: Fuhn (1970)

Panaspis africana (Gray 1845)   *Mocoa*
    Distribution: Principe, Rolas and Sao Tome Islands, Gulf of Guinea

Panaspis aloysiisabaudiae (Peracca 1907)   *Lygosoma*
    Distribution: Uganda

Panaspis amieti Perret 1973
    Distribution: Cameroon

Panaspis annobonensis Fuhn 1972
    Distribution: Annobon Island, Gulf of Guinea

Panaspis blochmanni (Tornier 1903)   *Lygosoma*
    Distribution: Zaire

Panaspis breviceps (Peters 1873)   *Euprepes*
    Distribution: Fernando Po, Gabon, Cameroon, Angola and Zaire

Panaspis burgeoni (Witte 1933)   *Lygosoma*
    Distribution: Ruwenzori Mountains, Zaire

Panaspis cabindae (Bocage 1866)   *Ablepharus*
    Distribution: Zaire south into Angola

Panaspis dewittei (Loveridge 1934)   *Siaphos*
   Distribution: Zaire

Panaspis duruarum (Monard 1949)   *Ablepharus*
   Distribution: Cameroon

Panaspis fuhni Perret 1973
   Distribution: Cameroon

Panaspis gemmiventris (Sjostedt 1897)   *Lygosoma*
   Distribution: Cameroon and Fernando Po

Panaspis graueri (Sternfeld 1912)   *Lygosoma*
   Distribution: Uganda, Rwanda, Burundi and Zaire

Panaspis hackarsi (Witte 1941)   *Lygosoma*
   Distribution: Zaire

Panaspis helleri (Loveridge 1932)   *Siaphos*
   Distribution: Ruwenzori Mountains, Zaire

Panaspis kilimensis (Stejneger 1891)   *Lygosoma*
   Distribution: Southern Sudan, Kenya, Tanzania, Uganda, Zaire and
              northern Angola

Panaspis lepesmei (Angel 1940)   *Lygosoma*
   Distribution: Cameroon

Panaspis luberoensis (Witte 1933)   *Lygosoma*
   Distribution: Zaire

Panaspis megalurus (Nieden 1913)   *Ablepharus*
   Distribution: Central Tanzania

Panaspis meleagris (Boulenger 1907)   *Lygosoma*
   Distribution: Uganda, Rwanda, Burundi and Zaire

Panaspis nimbensis (Angel 1944)   *Lygosoma*
   Distribution: Guinea and Ivory Coast
   Reference: Barbault (1975f) (1975)

Panaspis pauliani (Angel 1940)   *Lygosoma*
   Distribution: Cameroon

Panaspis quattuordigitata (Sternfeld 1912)   *Lygosoma*
   Distribution: Zaire

Panaspis reichenowi (Peters 1874)   *Lygosoma*
   Distribution: Cameroon, Congo, Zaire, Gabon and FErnando Po

Panaspis rhodurus (Laurent 1951)   *Lygosoma*
   Distribution: Zaire

Panaspis rohdei (Muller 1910)   *Lygosoma*
   Distribution: Cameroon and Gabon

Panaspis seydeli (Witte 1933)   *Ablepharus*
   Distribution: Zaire and Zambia
   Reference: Broadley (1971e)

Panaspis tancredii (Boulenger 1909)   *Ablepharus*
   Distribution: Ehtiopia

Panaspis thomasi (Tornier 1904)   *Lygosoma*
   Distribution: Kenya

Panaspis togoensis Werner 1902
   Distribution: Ivory Coast
   Reference:
   Panaspis kitsoni (Boulenger): Barbault (1974a)

Panaspis vigintiserierum (Sjostedt 1897)   *Lygosoma*
   Distribution: Cameroon and Fernando Po

Panaspis wahlbergi (Smith 1849)   *Cryptoblepharus*
   Distribution: Ethiopia and Congo south to Namibia, Botswana and
                Swaziland
   Reference: Fitzsimons (1943); Lanza and Carfi (1968); Parker (1942)

Panaspis wilsoni (Werner 1914)   *Ablepharus*
   Distribution: Sudan

Chapter 9

FAMILY LACERTIDAE

Order: Sauria
Suborder: Autarchoglossa
Infraorder: Scincomorpha
Superfamily: Lacertoidea
Family: Lacertidae

Genus: ACANTHODACTYLUS Wiegmann 1834
Species typica: *boskiana* Daudin
Reference: Boulenger (1921); Pasteur and Bons (1960)

Acanthodactylus boskianus asper (Audouin 1829)   *Lacerta*
    Distribution: Mauritania, Morocco east to Egypt south to Ethiopia,
        Sudan and Chad
    Reference: Angel and Lhote (1938); Duvdevani (1972); Duvdevani and
        Borut (1974a) (1974b)

Acanthodactylus boueti Chabanaud 1917
    Distribution: Benin

Acanthodactylus erythrurus atlanticus Boulenger 1918
    Distribution: Atlas Mountains of Morocco

Acanthodactylus erythrurus bellii Gray 1845
    Distribution: Algeria and Morocco
    Reference:
    Acanthodactylus vulgaris mauritanicus Doumergue: Boulenger (1921)

Acanthodactylus erythrurus lineomaculatus Dumeril and Bibron 1839
Distribution: Western and northwest Morocco

Acanthodactylus pardalis (Lichtenstein 1823)   *Lacerta*
Distribution: Morocco east to Egypt; Sudan
Reference: Bons (1960b); Busack (1975); Duvdevani (1972); Duvdevani
and Borut (1974a) (1974b)
Acanthodactylus pardalis bedriagae Lataste: Boulenger (1921)
Acanthodactylus pardalis latastii Boulenger: Boulenger (1921)
Acanthodactylus pardalis maculatus (Gray): Boulenger (1921)

Acanthodactylus savignyi savignyi (Audouin 1829)   *Lacerta*
Distribution: Morocco and Algeria
Reference: Bons (1968a)

Acanthodactylus savignyi blanci Doumergue 1901
Distribution: Tunisia

Acanthodactylus scutellatus scutellatus (Audouin 1829)   *Lacerta*
Distribution: Algerian Sahara east to Egypt
Reference: Bons and Girot (1963); Duvdevani (1972); Duvdevani and
Borut (1974a) (1974b)
Acanthodactylus scutellatus audouini Boulenger: Boulenger (1921)
Acanthodactylus scutellatus longipęs Boulenger: Boulenger (1921)

Acanthodactylus scutellatus aureus Gunther 1903
Distribution: Southern Morocco, Western Sahara and Mauritania

Acanthodactylus scutellatus dumerilii (Milne-Edwards 1829)   *Lacerta*
Distribution: Mauritania and Senegal

Acanthodactylus scutellatus inornatus (Gray 1838)   *Scapteira*
Distribution: Tunisia, Algeria and Morocco

Acanthodactylus spinicauda Doumergue 1901
Distribution: Algeria

Genus: ADOLFUS Sternfeld 1912
Species typica: *africanus* Boulenger
Reference: Arnold (1973); Boulenger (1920); Loveridge (1957)

Adolfus africanus (Boulenger 1906)   *Algiroides*
Distribution: Uganda, Zaire, Congo and Cameroon

Adolfus alleni (Barbour 1914)   *Algiroides*
    Distribution: Kenya and Uganda
    Reference: Andren (1976)

Adolfus vauereselli (Tornier 1902)   *Lacerta*
    Distribution: Congo, Zaire, Rwanda, Burundi, western Uganda and
            western Tanzania

Genus: APOROSAURA Boulenger 1887
Species typica: *anchietae* Bocage

Aporosaura anchietae (Bocage 1867)   *Pachyrhynchus*
    Distribution: Coastal Namibia and Angola
    Reference: Boulenger (1921); Brain (1962c); Fitzsimons (1943);
            Goldberg and Robinson (1979); Louw and Holm (1972);
            Robinson and Cunningham (1978)

Genus: BEDRIAGAIA Boulenger 1916
Species typica: *tropidopholis* Boulenger

Bedriagaia moreaui Loveridge 1936
    Distribution: Usambara Mountains, Tanzania

Bedriagaia tropidopholis Boulenger 1916
    Distribution: Zaire
    Reference: Boulenger (1921)

Genus: GALLOTIA Boulenger 1916
Species typica: *galloti* Dumeril and Bibron
Reference: Arnold (1973); Boulenger (1920)

Gallotia atlantica (Peters and Doria 1882)   *Lacerta*
    Distribution: Canary Islands: Lanzarote and Fuertaventura

Galllotia galloti galloti (Dumeril and Bibron 1839)   *Lacerta*
    Distribution: Canary Islands: Tenerife

Gallotia galloti caesaris (Lehrs 1914)   *Lacerta*
    Distribution: Canary Islands: Hierro

Gallotia galloti gomerae (Boettger and Muller 1914)   *Lacerta*
    Distribution: Canary Islands: Gomera

Gallotia galloti palmae (Boettger and Muller 1914)   *Lacerta*
    Distribution: Canary Islands: Las Palma

Gallotia simonyi (Steindachner 1889)   *Lacerta*
    Distribution: Roques del Zalmor, Hierro, now thought to be extinct

Gallotia stehlini (Schenkel 1901)   *Lacerta*
    Distribution: Canary Islands: Gran Canaria

Genus: GASTROPHOLIS Fischer 1886
Species typica: *vittatus* Fischer

Gastropholis vittatus Fischer 1886
    Distribution: Tanzania
    Reference: Boulenger (1921)

Genus: HOLASPIS Gray 1863
Species typica: *guentheri* Gray

Holaspis guentheri Gray 1863
    Distribution: Sierra Leone east to Uganda, Tanzania, Angola and
                  northern Mozambique
    Reference: Boulenger (1921); Laurent (1964b)

Genus: ICHNOTROPIS Peters 1854
Species typica: *capensis* Smith
Reference: Boulenger (1921)

Ichnotropis bivittata bivittata Bocage 1866
    Distribution: Angola, Zaire and Congo
    Reference: Broadley (1967b); Laurent (1964b); Loveridge (1957)

Ichnotropis bivittata pallida Laurent 1964
    Distribution: Angola

Ichnotropis capensis (Smith 1838)   *Algyra*
    Distribution: South Africa north to Tanzania and southern Zaire
    Reference: Broadley (1967b); Fitzsimons (1943); Loveridge (1953)
    Ichnotropis capensis longipes Boulenger: Fitzsimons (1943)
    Ichnotropis capensis overlaeti Witte and Laurent: Laurent (1964b)
    Ichnotropis longipes Boulenger: Boulenger (1921)

Ichnotropis chapini Schmidt 1919
    Distribution: Zaire

Ichnotropis grandiceps Broadley 1967
    Distribution: Botswana

Ichnotropis microlepidota Marx 1956
    Distribution: Angola

Ichnotropis squamulosa Peters 1854
    Distribution: Southern Angola, Namibia, Botswana, northern South
                Africa, Zimbabwe, Zambia, Mozambique and Tanzania
    Reference: Fitzsimons (1943); Huey and Pianka (1977); Loveridge
                (1953) (1957)

Ichnotropis tanganicana Boulenger 1917
    Distribution: Eastern shores of Lake Tanganyika, Tanzania

Genus: LACERTA Linnaeus 1758
Species typica: *agilis* Linnaeus
Reference: Arnold (1973); Boulenger (1920)

Lacerta andreanszkyi Werner 1929
    Distribution: Haut Atlas, Morocco

Lacerta australis Hewitt 1926
    Distribution: Cape Province
    Reference: Fitzsimons (1943)

Lacerta dugesii Milne-Edwards 1829
    Distribution: Madeira
    Reference: Crisp, Cook and Hereward (1979)

Lacerta echinata Cope 1862
    Distribution: Liberia east to northern Zaire

Lacerta jacksoni Boulenger 1899
Distribution: Tanzania, Kenya, Uganda west into northen Zaire

Lacerta lepida pater Lataste 1880
Distribution: Western Sahara, Morocco, Algeria and Tunisia
Reference: Arnold and Burton (1978); Pasteur and Bons (1960); Richter
(1933); Schmidt and Inger (1957); Sukhanov (1961);
Valverde (1957); Vogel (1964)

Lacerta perspicillata Dumeril and Bibron 1839
Distribution: Atlas Mountains of Morocco and Algeria

Lacerta rupicola Fitzsimons 1933
Distribution: Northern Transvaal
Reference: Fitzsimons (1943)

Genus: LATASTIA Bedriaga 1884
Species typica: *boscai* Bedriaga
Reference: Boulenger (1921); Parker (1942)

Latastia boscai boscai Bedriaga 1884
Distribution: Ethiopia and northwestern Somalia
Reference:
Latastia wachei Werner: Boulenger (1921)

Latastia boscai arenicola Parker 1942
Distribution: Central Somalia

Latastia boscai burii Boulenger 1907
Distribution: Northeastern Somalia

Latastia carinata (Peters 1874)   *Lacerta*
Distribution: Somalia

Latastia johnstonii Boulenger 1907
Distribution: Malawi, Zambia and Tanzania
Reference: Broadley (1971e); Loveridge (1953) (1957)

Latastia longicaudata longicaudata (Reuss 1834)   *Lacerta*
Distribution: Senegal east through northern Nigeria to Sudan; Ethiopia
and Egypt

Latastia longicaudata caeruleopunctata Parker 1935
Distribution: North and west Haud, Somalia

Latastia longicaudata doriai Bedriaga 1884
Distribution: Northwestern Somalia

Latastia longicaudata revoili (Vaillant 1882)   *Eremias*
Distribution: Southern Ethiopia, Somalia and Kenya
Reference: Loveridge: (1957)

Latastia siebenrocki (Tornier 1905)   *Eremias*
Distribution: Guinea east to Tanzania

Latastia taylori Parker 1942
Distribution: Somalia

Genus: MEROLES Gray 1838
Species typica: *knoxii* Milne-Edwards
Reference: Boulenger (1921); Fitzsimons (1943)

Meroles ctenodactylus (Smith 1838)   *Lacerta*
Distribution: Namibia and adjacent Cape Province

Meroles cuneirostris (Strauch 1867)   *Podarces*
Distribution: Coastal Namibia
Reference: Brain (1962c); Goldberg and Robinson (1979); Robinson
         and Cunningham (1978)

Meroles knoxii (Milne-Edwards 1829)   *Lacerta*
Distribuion: Western Cape Province and Namibia
Reference: Haacke (1965)
Scaptira knoxii pequensis Hewitt: Fitzsimons (1943)

Meroles micropholidotus (Mertens 1937)   *Scapteira*
Distribution: Namibia
Reference: Mitchell and Steyn (1967)

Meroles reticulatus (Bocage 1867)   *Scapteira*
Distribution: Coastal Namibia and Angola

Meroles suborbitalis (Peters 1869)   *Eremias*
Distribuion: Northwestern Cape Province and Namibia
Reference: Brain (1962c); Huey and Pianka (1977); Stebbins (1961)

Genus: MESALINA Gray 1838
Species typica: *rubropunctata* Lichtenstein
Reference: Boulenger (1921); Fitzsimons (1943); Szczerbak (1974)

Mesalina benguelensis (Bocage 1867)   *Eremias*
    Distribution: Angola

Mesalina brenneri (Peters 1869)   *Eremias*
    Distribution: Somalia
    Reference: Parker (1942)

Mesalina burchelli (Dumeril and Bibron 1839)   *Eremias*
    Distribution: Cape Province east to Lesotho

Mesalina capensis (Smith 1838)   *Lacerta*
    Distribution: Namibia south to Cape Province

Mesalina ercolinii (Lanza and Poggesi 1975)   *Eremias*
    Distribution: Central Somalia

Mesalina erythrosticta (Boulenger 1891)   *Eremias*
    Distribution: Somalia
    Reference: Parker (1942)

Mesalina guineensis (Boulenger 1887)   *Eremias*
    Distribution: Nigeria

Mesalina guttulata guttulata (Lichtenstein 1823)   *Lacerta*
    Distribution: Morocco east to Egypt
    Reference: Haas (1951); Pasteur and Bons (1960)

Mesalina lineoocellata lineoocellata (Dumeril and Bibron 1837)
                                                                              Eremias
    Distribution: Botswana, Namibia and adjacent South Africa
    Reference: Huey and Pianka (1977); Stebbins (1961)
    Eremias aspera Boulenger: Boulenger (1921)

Mesalina lineoocellata pulchella (Gray 1845)   *Eremias*
    Distribution: Namibia and Cape Province

Mesalina lugubris (Smith 1838)   *Lacerta*
    Distribution: Namibia, Botswana, southern Angola, Zimbabwe and
                Transvaal
    Reference: Huey and Pianka (1977)

Mesalina mucronata (Blanford 1870)   *Acanthodactylus*
    Distribution: Egypt, Sudan, Ethiopia and northern Somalia
    Reference: Parker (1942)

Mesalina namaquensis (Dumeril and Bibron 1839)   *Eremias*
    Distribution: Southern Angola, Namibia, Botswana and northern Cape
        Province
    Reference: Huey and Pianka (1977)

Mesalina neumanni (Tornier 1905)   *Eremias*
    Distribution: Ethiopia and Kenya

Mesalina nitida nitida (Gunther 1872)   *Eremias*
    Distribution: Togo, Benin and Nigeria

Mesalina nitida garambensis (Schmidt 1919)   *Eremias*
    Distribution: Zaire

Mesalina olivieri olivieri (Audouin 1829)   *Lacerta*
    Distribution: Western Sahara and Morocco east to Egypt, south (west of
        the Nile) to Sudan, Ethiopia and Somalia
    Reference: Boulenger (1921); Haas (1951); Parker (1942); Pasteur and
        Bons (1960)

Mesalina olivieri latasti (Boulenger 1918)   *Eremias*
    Distribution: Vicinity of Aumale, Algeria
    Reference: Haas (1951)

Mesalina olivieri martini (Boulenger 1897)   *Eremias*
    Distribution: African coast of the Red Sea
    Reference: Haas (1951)

Mesalina olivieri simoni (Boettger 1881)   *Podarcis*
    Distribution: Atlantic side of the Haut Atlas, Morocco
    Reference: Pasteur and Bons (1960)

Mesalina olivieri susana (Boulenger 1918)   *Eremias*
    Distribution: Vicinity of Susa, Tunisia
    Reference: Haas (1951)

Mesalina pasteuri (Bons 1960)   *Eremias*
    Distribution: Morocco

Mesalina quadrinasalis (Chabanaud 1918)   *Eremias*
    Distribution: Chad

Mesalina rubropunctata (Lichtenstein 1823)   *Lacerta*
    Distribution: Morocco east to Egypt
    Reference: Angel and Lhote (1938); Gauthier (1966); Pasteur and Bons
        (1960)

Mesalina septemstriata (Parker 1942)   *Eremias*
    Distribution: Somalia

Mesalina smithii (Boulenger 1895)   *Eremias*
    Distribution: Ethiopia, Somalia, Kenya and Tanzania
    Reference: Parker (1942)

Mesalina spekii spekii (Gunther 1872)   *Eremias*
    Distribution: Tanzania, Uganda and Kenya

Mesalina spekii sextaeniata (Stejneger 1894)   *Eremias*
    Distribution: Kenya, Somalia and Ethiopia
    Reference: Loveridge (1957); Parker (1942)

Mesalina striata (Peters 1874)   *Eremias*
    Distribution: Somalia
    Reference: Parker (1942)

Mesalina undata (Smith 1838)   *Lacerta*
    Distribution: Angola, Namibia and northwestern Cape Province
    Reference: Laurent (1964b)

Genus: NUCRAS Gray 1838
Species typica: *lalandii* Milne-Edwards
Reference: Boulenger (1920); Broadley (1972a)

Nucras boulengeri boulengeri Neumann 1900
    Distribution: Northern Tanzania and adjacent Kenya

Nucras boulengeri kilosae Loveridge 1922
    Distribution: Tanzania

Nucras caesicaudata Broadley 1972
    Distribution: Gazaland Plain of Mozambique and Zimbabwe

Nucras emini Boulenger 1907
    Distribution: Northern Tanzania

Nucras intertexta (Smith 1838)   *Lacerta*
    Distribution: Mozambique, Zimbabwe, Namibia, Botswana and
                  South Africa
    Reference: Fitzsimons (1943); Huey and Pianka (1977)

Nucras lalandii (Milne-Edwards 1829)   *Lacerta*
    Distribution: South Africa
    Reference: Fitzsimons (1943)

Nucras scalaris Laurent 1964
    Distribution: Angola

Nucras taeniolata taeniolata (Smith 1838)   *Lacerta*
    Distribution: Albany District of Cape Province

Nucras taeniolata ornata (Gray 1864)   *Teira*
    Distribution: Zambia and Malawi south to Natal west to Botswana;
            northern Namibia
    Reference:
    Nucras intertexta damarana Parker: Fitzsimons (1943)
    Nucras intertexta holubi (Steindachner): Boulenger (1920); Fitzsimons
        (1943)

Nucras tessellata tessellata (Smith 1838)   *Lacerta*
    Distribution: Angola, Namibia and western Cape Province
    Reference: Fitzsimons (1943); Huey and Pianka (1977); Laurent (1964b)

Nucras tessellata livida (Smith 1838)   *Lacerta*
    Distribution: Central and southern Cape Province

Genus: OPHISOPS Menetries 1832
Species typica: *elegans* Menetries
Reference: Boulenger (1921)

Ophisops elbaensis Schmidt and Inger 1957
    Distribution: Northeastern Sudan

Ophisops elegans elegans Menetries 1832
    Distribution: Egypt, Libya and northern Sudan

Ophisops elegans occidentalis Boulenger 1887
    Distribution: Morocco, Algeria, Tunisia and northern Libya

Genus: PHILOCHORTUS Matschie 1893
Species typica: *neumanni* Matschie
Reference: Angel (1936); Boulenger (1921); Parker (1942)

Philochortus hardeggeri hardeggeri (Steindachner 1891)   *Latastia*
  Distribution: Coastal northern Somalia

Philochortus hardeggeri heterolepis (Boettger 1893)   *Eremias*
  Distribution: Northern Somalia

Philochortus intermedius intermedius Boulenger 1917
  Distribution: Northern Kenya, Somalia, Sudan, Ethiopia and Egypt
  Reference: Marx (1968)

Philochortus intermedius rudolfensis Parker 1932
  Distribution: near the mouth of the Kaliokwell River, Kenya

Philochortus lhotei Angel 1936
  Distribution: Niger and southern Algeria

Philochortus phillipsii (Boulenger 1898)   *Latastia*
  Distribution: Somalia

Philochortus spinalis (Peters 1874)   *Lacerta*
  Distribution: Ethiopia west to Niger

Philochortus zolli Scortecci 1934
  Distribution: Southern Libya

Genus: PODARCIS Wagler 1830
Species typica: *muralis* Laurenti
Reference: Arnold (1973)

Podarcis hispanica vaucheri (Boulenger 1905)   *Lacerta*
  Distribution: Morocco, Algeria and Tunisia
  Reference: Arnold and Burton (1978); Pasteur and Bons (1960)
  Podarcis hispanica bocagii Seoane: Boulenger (1920, part)

Genus: POROMERA Boulenger 1887
Species typica: *fordii* Hallowell

Poromera fordii (Hallowell 1857)   *Tachydromus*
  Distribution: Gabon, Cameroon and Congo
  Reference: Boulenger (1921)

Genus: PSAMMODROMUS Fitzinger 1826
Species typica: *hispanicus* Fitzinger
Reference: Arnold (1973); Boulenger (1921); Pasteur and Bons (1960)

Psammodromus algirus algirus (Linnaeus 1758)   *Lacerta*
    Distribution: Mauritania, Western Sahara, Morocco and Algeria
    Reference: Arnold and Burton (1978); Schmidt and Inger (1957)

Psammodromus algirus nollii (Fischer 1887)   *Tropidosaura*
    Distribution: Northern Morocco, Algeria and Tunisia

Psammodromus blanci (Lataste 1880)   *Zerzoumia*
    Distribution: Northeastern Morocco and Algeria

Psammodromus microdactylus (Boettger 1881)   *Algira*
    Distribution: Morocco

Genus: TROPIDOSAURA Fitzinger 1826
Species typica: *montana* Gray
Reference: Fitzsimons (1943)

Tropidosaura cottrelli (Hewitt 1925)   *Basutosaura*
    Distribution: Lesotho

Tropidosaura essexi Hewitt 1927
    Distribution: Lesotho

Tropidosaura gularis Hewitt 1927
    Distribution: Western and southwestern Cape Province

Tropidosaura montana montana (Gray 1831)   *Lacerta*
    Distribution: Western and southwestern Cape Province

Tropidosaura montana rangeri Hewit 1926
    Distribution: Southeastern and eastern Cape Province

Chapter 10

FAMILY CORDYLIDAE

Order: Sauria
Suborder: Autarchoglossa
Infraorder: Scincomorpha
Superfamily: Lacertoidea
Family: Cordylidae
Subfamily: Gerrhosaurinae

Genus: ANGOLOSAURUS Fitzsimons 1953
Species typica: *skoogi* Andersson

Angolosaurus skoogi (Andersson 1916)   *Gerrhosaurus*
    Distribution: Namibia and southern Angola
    Reference: Hamilton and Coetzee (1969); Steyn (1963)

Genus: CORDYLOSAURUS Gray 1865
Species typica: *subtessellatus* Smith

Cordylosaurus subtessellatus (Smith 1844)   *Gerrhosaurus*
    Distribution: Western Cape Province north to southern Angola
    Reference: Fitzsimons (1943); Loveridge (1942b)
    Cordylosaurus subtessellatus australis Hewitt: Fitzsimons (1943)
    Cordylosaurus subtessellatus trivittatus (Peters): Fitzsimons (1943)

Genus: GERRHOSAURUS Wiegmann 1828
Species typica: *flavigularis* Wiegmann
Reference: Fitzsimons (1943); Loveridge (1942b)

Gerrhosaurus flavigularis flavigularis Wiegmann 1828
   Distribution: Malawi, Mozambique, eastern South Africa, Swaziland,
         Zambia and Zimbabwe
   Reference: Gabe and Saint Girons (1969); Mitchell and Steyn (1965)

Gerrhosaurus flavigularis fitzsimonsi Loveridge 1942
   Distribution: Ethiopia, Kenya, Somalia, eastern Sudan and Tanzania

Gerrhosaurus major major Dumeril 1851
   Distribution: Eastern Kenya, Tanzania and Mozambique
   Reference: Schmidt and Inger (1957)

Gerrhosaurus major bottegoi Prato 1895
   Distribution: Ethiopia, Kenya, Somalia and northern Tanzania
   Reference: Loveridge (1957); Parker (1942)

Gerrhosaurus major grandis Boulenger 1908
   Distribution: Malawi, southern Mozambique, Swaziland, Transvaal
         and Zimbabwe

Gerrhosaurus major zechi Tornier 1901
   Distribution: Southern Sudan and northern Zaire west to Togo
   Reference: Camp (1923)

Gerrhosaurus multilineatus multilineatus Bocage 1866
   Distribution: Angola, Zaire and Zambia
   Reference: Broadley (1971e)
   Gerrhosaurus auritus bulsi Laurent: Laurent (1954) (1964b)
   Gerrhosaurus nigrolineatus anselli Broadley: Broadley (1960b)

Gerrhosaurus multilineatus auritus Boettger 1887
   Distribution: Namibia and Botswana
   Reference: Broadley (1971e)

Gerrhosaurus nigrolineatus Hallowell 1857
   Distribution: Gabon, Congo, Zaire, Angola, Namibia, Botswana,
         Transvaal, Zimbabwe, Malawi, Mozambique, Tanzania,
         Kenya and Zambia

Gerrhosaurus typicus (Smith 1836)   *Pleurotuchus*
   Distribution: Little Namaqualand

Gerrhosaurus validus validus Smith 1849
   Distribution: Transvaal, southern Mozambique, Swaziland and
         Zimbabwe

Gerrhosaurus validus maltzahni de Grys 1938
   Distribution: Namibia and southern Angola

Genus: TETRADACTYLUS Merrem 1820
Species typica: *tetradactyla* Lacepede
Reference: Fitzsimons (1943); Loveridge (1942b)

Tetradactylus africanus (Gray 1838)   *Caitia*
   Distribution: South Africa

Tetradactylus eastwoodae Methuen and Hewitt 1913
   Distribution: Transvaal

Tetradactylus ellenbergeri (Angel 1922)   *Paratetradactylus*
   Distribution: Angola, Zambia and Zaire
   Reference: Broadley (1971e); Laurent (1964b)
   Tetradactylus ellenbergeri boulengeri Witte: Laurent (1964b);
                                         Loveridge (1942b)
   Tetradactylus fitzsimonsi simplex Laurent: Loveridge (1957)

Tetradactylus fitzsimonsi Hewitt 1915
   Distribution: Cape Province north to Zimbabwe

Tetradactylus seps seps (Linnaeus 1758)   *Lacerta*
   Distribution: Southwestern Cape Province

Tetradactylus seps laevicauda Hewitt 1915
   Distribution: Cape Province, Orange Free State and Natal

Tetradactylus tetradactylus tetradactylus (Lacepede 1803)   *Lacerta*
   Distribution: Southwestern Cape Province

Tetradactylus tetradactylus bilineatus Hewitt 1926
   Distribution: Burghersdorp, Cape Province

Genus: TRACHELOPTYCHUS Peters 1854
Species typica: *madagascariensis* Peters
Reference: Angel (1942)

Tracheloptychus madagascariensis Peters 1854
   Distribution: Madagascar

Tracheloptychus petersi Grandidier 1869
   Distribution: Madagascar

Genus: ZONOSAURUS Boulenger 1887
Species typica: *madagascariensis* Gray
Reference: Angel (1942)

Zonosaurus aeneus (Grandidier 1872)   *Gerrhosaurus*
   Distribution: Madagascar

Zonosaurus boettgeri Steindachner 1891
   Distribution: Madagascar

Zonsaurus karsteni (Grandidier 1869)   *Gerrhosaurus*
   Distribution: Madagascar

Zonosaurus laticaudatus (Grandidier 1869)   *Gerrhosaurus*
   Distribution: Madagascar

Zonosaurus madagascariensis (Gray 1845)   *Cicigna*
   Distribution: Madagascar and Aldabra

Zonosaurus maximus Boulenger 1896
   Distribution: Madagascar

Zonosaurus ornatus (Gray 1845)   *Cicigna*
   Distribution: Madagascar

Zonosaurus quadrilineatus (Grandidier 1867)   *Gerrhosaurus*
   Distribution: Madagascar

Zonosaurus rufipes rufipes (Boettger 1881)   *Gerrhosaurus*
   Distribution: Madagascar

Zonosaurus rufipes subunicolor (Boettger 1881)   *Gerrhosaurus*
   Distribution: Madagascar

Zonosaurus trilineatus Angel 1939
   Distribution: Madagascar

Subfamily: Cordylinae

Genus: CHAMAESAURA Schneider 1799
Species typica: *anguina* Linnaeus
Reference: Fitzsimons (1943)

Chamaesaura aenea (Wiegmann 1843)   *Cricochalcis*
    Distribution: Transvaal south to eastern Cape Province

Chamaesaura anguina anguina (Linnaeus 1758)   *Lacerta*
    Distribution: Cape Province north to Transvaal and southern
                Mozambique
    Reference: Plessis (1945); Zavattari (1910)

Chamaesaura anguina oligopholis Laurent 1964
    Distribution: Angola

Chamaesaura anguina tenuior Gunther 1895
    Distribution: Eastern Zaire, Uganda, Kenya and Tanzania

Chamaesaura macrolepis macrolepis (Cope 1862)   *Mancus*
    Distribution: Swaziland and adjacent South Africa

Chamaesaura macrolepis miopropus Boulenger 1894
    Distribution: Tanzania west through Zambia to Angola
    Reference: Broadley (1971e); Loveridge (1957)

Genus: CORDYLUS Laurenti 1768
Species typica: *cordylus* Linnaeus
Reference: Fitzsimons (1943); Loveridge (1944b)

Cordylus campbelli Fitzsimons 1938
    Distribution: Namibia

Cordylus cataphractus Boie 1828
    Distribution: Western Cape Province
    Reference: Schmidt and Inger (1957)

Cordylus coeruleopunctatus (Methuen and Hewitt 1913)   *Zonurus*
    Distribution: Attaqua and Outeniqua Mountains of South Africa

Cordylus cordylus cordylus (Linnaeus 1758)   *Lacerta*
    Distribution: Southern Cape Province
    Reference: Burrage (1974); Gabe and Saint Girons (1969) (1972);
            Versluys (1898)

Cordylus cordylus angolensis (Bocage 1895)   *Zonurus*
    Distribution: Angola

Cordylus cordylus jonesii (Boulenger 1891)   *Zonurus*
    Distribution: Transvaal, Botswana, Zimbabwe, Swaziland, southern
            Mozambique, southern Angola and southwestern Zambia

Cordylus cordylus lawrenci (Fitzsimons 1939)   *Zonurus*
    Distribution: Cape Province

Cordylus cordylus minor Fitzsimons 1943
    Distribution: Cape Province

Cordylus cordylus niger Cuvier 1829
    Distribution: South Africa

Cordylus cordylus pustulatus (Peters 1862)   *Zonurus*
    Distribution: Otjimbingwe, South Africa

Cordylus cordylus rhodesianus (Hewitt 1933)   *Zonurus*
    Distribution: Northeastern Zimbabwe

Cordylus cordylus tasmani (Power 1930)   *Zonurus*
    Distribution: Eastern Cape Province

Cordylus cordylus tropidosternum (Cope 1869)   *Zonurus*
    Distribution: Kenya, Tanzania, Mozambique, eastern Zambia, eastern
            Zimbabwe and Malawi

Cordylus giganteus Smith 1844
    Distribution: South Africa
    Reference: Camp (1923); Lakjer (1926); Richter (1933)

Cordylus macropholis (Boulenger 1910)   *Zonurus*
    Distribution: Coastal Little Namaqualand
    Reference: McLachlan and Spence (1966a)

Cordylus namaquensis (Methuen and Hewitt 1914)   *Zonurus*
    Distribution: Namibia

Cordylus peersi (Hewitt 1932)   *Zonurus*
    Distribution: Little Namaqualand

Cordylus polyzonus polyzonus Smith 1838
    Distribution: Cape Province, western Orange Free State and southern
            Namibia
    Reference: Richter (1933)

Cordylus polyzonus jordani (Parker 1936)   *Zonurus*
    Distribution: Namibia
    Reference: Steyn (1962)

Cordylus ukingensis (Loveridge 1932)   *Zonurus*
    Distribution: Southern highlands of Tanzania

Cordylus vittifer vittifer (Reichenow 1887)   *Zonurus*
    Distribution: Transvaal, Swaziland, Natal, Zululand and Botswana
    Reference: Jacobsen (1972)

Cordylus vittifer machadoi Laurent 1964
    Distribution: Angola

Cordylus warreni warreni (Boulenger 1908)   *Zonurus*
    Distribution: Ubombo Mountains, Zululand

Cordylus warreni barbertonensis (Van Dam 1921)   *Zonurus*
    Distribution: Eastern Transvaal and Swaziland

Cordylus warreni breyeri (Van Dam 1921)   *Zonurus*
    Distribution: Waterberg District, Transvaal

Cordylus warreni depressus (Fitzsimons 1930)   *Zonurus*
    Distribution: North Transvaal

Cordylus warreni laevigatus (Fitzsimons 1933)   *Zonurus*
    Distribution: Zoutpansberg Mountains, Transvaal

Cordylus warreni mossambicus Fitzsimons 1958
    Distribution: Chimanimani Mountains of Zimbabwe-Mozambique

Cordylus warreni perkoensis (Fitzsimons 1930)   *Zonurus*
    Distribution: Eastern Transvaal south of Olifants River

Cordylus warreni regius Broadley 1962
    Distribution: Southern Zimbabwe

Cordylus warreni vandami (Fitzsimons 1930)   *Zonurus*
    Distribution: Eastern Transvaal north of Olifants River

Genus: PLATYSAURUS Smith 1844
Species typica: *capensis* Smith
Reference: Broadley (1978); Fitzsimons (1943); Loveridge (1944b)

Platysaurus capensis Smith 1844
    Distribution: Southern Namibia and adjacent Cape Province
    Reference: Haacke (1965)

Platysaurus fitzsimonsi Loveridge 1944
    Distribution: Sekukuni Mountains, Transvaal

Platysaurus guttatus guttatus Smith 1849
    Distribution: Northern and eastern Transvaal
    Reference:
    Platysaurus minor orientalis Fitzsimons: Fitzsimons (1943)

Platysaurus guttatus minor Fitzsimons 1930
    Distribution: Waterberg, Transvaal

Platysaurus imperator Broadley 1962
    Distribution: Northeastern Zimbabwe and adjacent Mozambique

Platysaurus intermedius intermedius Matschie 1891
    Distribution: Eastern Transvaal

Platysaurus intermedius natalensis Fitzsimons 1948
    Distribution: Natal and southern Swaziland

Platysaurus intermedius nyase Loveridge 1953
    Distribution: Middle Shire Valley, Malawi west into adjacent
                  Mozambique

Platysaurus intermedius parvus Broadley 1976
    Distribution: Blouberg, Northern Transvaal

Platysaurus intermedius rhodesianus Fitzsimons 1941
    Distribution: Zimbabwe, eastern Botswana, Transvaal north of
                  Zoutpansberg and the southern part of the Manica
                  Platform, Mozambique

Platysaurus intermedius subniger Broadley 1962
  Distribution: Northeastern Zimbabwe and adjacent Mozambique

Platysaurus intermedius wilhelmi Hewitt 1909
  Distribution: Southeastern Transvaal, Swaziland, Zululand and adjacent
    Mozambique

Platysaurus maculatus maculatus Broadley 1965
  Distribution: Mozambique north of the Zambesi

Platysaurus maculatus lineicauda Broadley 1965
  Distribution: Mozambique north of the Zambesi and east of the Shire

Platysaurus mitchelli Loveridge 1953
  Distribution: Mlanje Mountain, Malawi

Platysaurus ocellatus Broadley 1962
  Distribution: Chimanimani Mountains of Zimbabwe-Mozambique

Platysaurus pungweensis pungweensis Broadley 1959
  Distribution: Eastern Zimbabwe and adjacent Mozambique

Platysaurus pungweensis blakei Broadley 1964
  Distribution: Southern Manica Platform, Mozambique

Platysaurus relictus Broadley 1976
  Distribution: Transvaal

Platysaurus torquatus Peters 1879
  Distribution: Southern Malawi, west-central Mozambique and
    northeastern Zimbabwe

Genus: PSEUDOCORDYLUS Smith 1838
Species typica: *microlepidotus* Cuvier
Reference: Fitzsimons (1943); Loveridge (1944b)

Pseudocordylus capensis (Smith 1838)   *Cordylus*
  Distribution: Cape Province from False Bay to Mossel Bay

Pseudocordylus langi Loveridge 1944
  Distribution: Lesotho west into Cape Province

Pseudocordylus microlepidotus microlepidotus (Cuvier 1829)   *Cordylus*
    Distribution: Western and southwestern Cape Province

Pseudocordylus microlepidotus fasciatus (Smith 1838)   *Cordylus*
    Distribution: East and east-central Cape Province

Pseudocordylus microlepidotus melanotus (Smith 1838)   *Cordylus*
    Distribution: Transvaal south to Natal, west through Orange Free State
            to Cape Province
    Reference:
    Pseudocordylus subviridis subviridis (Smith): Fitzsimons (1943)
    Pseudocordylus subviridis transvaalensis Fitzsimons: Fitzsimons
                            (1943)

Pseudocordylus microlepidotus namaquensis Hewitt 1927
    Distribution: Little Namaqualand

Pseudocordylus robertsi (Van Dam 1921)   *Zonurus*
    Distribution: Western Cape Province

Pseudocordylus spinosus Fitzsimons 1947
    Distribution: Natal

Chapter 11

FAMILY ANGUIDAE

Order: Sauria
Suborder: Autarchoglossa
Infraorder: Anguimorpha
Superfamily: Anguioidea
Family: Anguidae
Subfamily: Anguinae

Genus: ANGUIS Linnaeus 1758
Species typica: *fragilis* Linnaeus

Anguis fragilis fragilis Linnaeus 1758
    Distribution: Algeria and Tunisia
    Reference: Arnold and Burton (1978); Bannister (1968); Kirk and
           Hogben (1945); Trost (1953)

Genus: OPHISAURUS Daudin 1803
Species typica: *ventralis* Linnaeus

Ophisaurus koellikeri (Gunther 1873)   *Hyalosaurus*
    Distribution: North and western Morocco
    Reference: Bons and Bons (1969); Gabe and Saint Girons (1965) (1969);
           Pasteur and Bons (1960)

Chapter 12

FAMILY VARANIDAE

Order: Sauria
Suborder: Autarchoglossa
Infraorder: Anguimorpha
Superfamily: Varanoidea
Family: Varanidae

Genus: VARANUS Merrem 1820
Species typica: *nilotica* Linnaeus

Varanus exanthematicus exanthematicus (Bosc 1792)   *Lacerta*
    Distribution: Senegal east to Ethiopia
    Reference: Laurent (1964); Schmidt and Inger (1957)

Varanus exanthematicus albigularis (Daudin 1802)   *Tupinambis*
    Distribution: Namibia, Botswana, South Africa, Zimbabwe,
                Mozambique and Zambia
    Reference: Fitzsimons (1943)

Varanus exanthematicus angolensis Schmidt 1934
    Distribution: Angola

Varanus exanthematicus ionidesi Laurent 1964
    Distribution: Southeastern Tanzania

Varanus exanthematicus microstictus Boettger 1893
    Distribution: Southern Ethiopia, southern Somalia, Kenya, Tanzania,
                Uganda and Mozambique

Varanus griseus griseus (Daudin 1803)   *Tupinambis*
    Distribution: Western Sahara and Mauritania east to Sudan and Egypt
    Reference: Khalil and Abdel-Messeih (1959a) (1959b) (1959c) (1961b);
            Vogel (1964)

Varanus niloticus (Linnaeus 1766)   *Lacerta*
    Distribution: South Africa north to Egypt in the east and Liberia in
            the west
    Reference: Cloudsley-Thompson (1967) (1969); Cowles (1930);
            Fitzsimons (1943); Gabe (1972); Hirth and Latif (1979);
            Lakjer (1926); Mitchell and Steyn (1967); Richter (1933);
            Rieppel and Labhardt (1979); Wood and Johansen (1974)

Chapter 13

FAMILY LEPTOTYPHLOPIDAE

Order: Serpentes
Suborder: Scolecophidia
Family: Leptotyphlopidae

Genus: LEPTOTYPHLOPS Fitzinger 1843
Species typica: *nigricans* Schlegel

Leptotyphlops bicolor (Jan 1860)   *Stenostoma*
    Distribution: Togo, Benin, southwest Chad, Ghana, Nigeria, Upper
            Volta, Niger, Ivory Coast and Mali
    Reference: Branch (1979)

Leptotyphlops boulengeri (Boettger 1913)   *Glauconia*
    Distribution: Manda and Lamu Islands, Kenya

Leptotyphlops brevicaudus (Bocage 1887)   *Stenostoma*
    Distribution: Benin, Ghana, Ivory Coast, Mali, Niger and Nigeria

Leptotyphlops cairi (Dumeril and Bibron 1844)   *Stenostoma*
    Distribution: Libya, Egypt, Ethiopia, Sudan and Somalia with an
            isolated population in Niger and Mauritania
    Reference: Hahn (1978b)

Leptotyphlops conjunctus conjunctus (Jan 1861)   *Stenostoma*
    Distribution: Eastern Cape Province north to southern Transvaal,
            Swaziland and Natal
    Reference: Broadley and Watson (1976)

Leptotyphlops conjunctus incognitus Broadley and Watson 1976
    Distribution: Southern Zambia, southern Malawi, northern and eastern
            Zimbabwe, northern and eastern Transvaal, central and
            southern Mozambique, eastern Swaziland and KwaZulu;
            an isolated population in central Namibia

Leptotyphlops conjunctus latirostris (Sternfeld 1912)   *Glauconia*
    Distribution: Kivu Province, Zaire
    Reference: Laurent (1956a)

Leptotyphlops conjunctus lepezi (Boulenger 1901)   *Glauconia*
    Distribution: Southern Zaire and Zambia
    Reference: Laurent (1956a)

Leptotyphlops debilis (Chabanaud 1918)   *Glauconia*
    Distribution: Chad west through West Africa

Leptotyphlops dissimilis (Bocage 1886)   *Glauconia*
    Distribution: Sudan

Leptotyphlops distanti (Boulenger 1892)   *Glauconia*
    Distribution: Transvaal and southern Mozambique, relict populations
            in Natal and KwaZulu
    Reference: Broadley and Watson (1976); Fitzsimons (1962) (1974)

Leptotyphlops filiformis (Boulenger 1899)   *Glauconia*
    Distribution: Socotra

Leptotyphlops fitzingeri (Jan 1861)   *Stenostoma*
    Distribution: Isle of Rhodes near Cairo, Egypt

Leptotyphlops gestri (Boulenger 1906)   *Glauconia*
    Distribution: Central African Republic, Cameroon and Fernando Po

Leptotyphlops gracilior (Boulenger 1910)   *Glauconia*
    Distribution: Western Cape Province
    Reference: Broadley and Watson (1976); Fitzsimons (1974)

Leptotyphlops labialis (Sternfeld 1908)   *Glauconia*
    Distribution: Northern Namibia
    Reference: Broadley and Watson (1976); Fitzsimons (1962) (1974)

Leptotyphlops longicaudus (Peters 1854)   *Stenostoma*
    Distribution: Southern coastal Somalia south to eastern Transvaal
            and Swaziland, west to Zimbabwe and eastern Botswana
    Reference: Broadley and Watson (1976); Fitzsimons (1974)

Leptotyphlops macrorhynchus macrorhynchus (Jan 1861)   *Stenostoma*
    Distribution: Senegal, Mauritania, Mali, Ghana, Guinea, Morocco,
            Algeria, Libya, Tunisia, Egypt, Ethiopia, Sudan and
            Somalia
    Reference: Hahn (1978b)
    Leptotyphlops braccianii (Scortecci): Hahn (1978a); Parker (1949)
    Leptotyphlops erythraeus (Scortecci): Hahn (1978a); Parker (1949)

Leptotyphlops macrorhynchus bilmaensis Angel 1936
    Distribution: Niger

Leptotyphlops macrurus (Boulenger 1903)   *Glauconia*
    Distribution: Socotra

Leptotyphlops narirostris narirostris (Peters 1867)   *Stenostoma*
    Distribution: Nigeria, Cameroon, Benin, Togo, Guinea-Bissau, Ivory
            Coast and southern Niger

Leptotyphlops narirostris boueti (Chabanaud 1917)   *Glauconia*
    Distribution: Senegal, Mauritania and Mali

Leptotyphlops natatrix (Andersson 1937)   *Glauconia*
    Distribution: Tobo Swamp, Gambia

Leptotyphlops nigricans and nigricans (Schlegel 1839)   *Typhlops*
    Distribution: Southern Cape Province; central Zambia north through
            eastern Angola, eastern Zaire, Rwanda, Tanzania, Kenya,
            Uganda and southern Sudan
    Reference: Broadley and Watson (1976); Brock (1932); Fitzsimons
            (1974); Hahn (1978b)

Leptotyphlops nigricans pembae Loveridge 1941
    Distribution: Pemba Island, Tanzania
    Reference: Broadley and Watson (1976)

Leptotyphlops occidentalis Fitsimons 1962
    Distribution: Namibia
    Reference: Broadley and Watson (1976); Fitzsimons (1974)

Leptotyphlops reticulatus (Boulenger 1906)   *Glauconia*
    Distribution: Northwest Somalia
    Reference: Parker (1949)

Leptotyphlops rostratus (Bocage 1886)   *Stenostoma*
    Distribution: Angola

Leptotyphlops scutifrons scutifrons (Peters 1854) *Stenostoma*
  Distribution: Southern Tanzania south to Natal, west to Angola and
      Namibia
  Reference: Broadley and Watson (1976); Rieppel (1979)

Leptotyphlops scutifrons merkeri (Werner 1909) *Glauconia*
  Distribution: Tanzania and Kenya
  Reference: Broadley and Watson (1976)

Leptotyphlops signatus (Jan 1861) *Stenostoma*
  Distribution: Tabora, Tanzania and Kibwezi, Kenya

Leptotyphlops sundevalli (Jan 1861) *Stenostoma*
  Distribution: Ghana and Togo

Leptotyphlops telloi Broadley and Watson 1976
  Distribution: Lebombo Range on the Mozambique-Swaziland border

Leptotyphlops variabilis (Scortecci 1928) *Glauconia*
  Distribution: Eritrea, Ethiopia

Genus: RHINOLEPTUS Orejas-Miranda, Roux-Esteve and Guibe 1970
Species typica: *koniagui* Villiers

Rhinoleptus koniagui (Villiers 1956) *Typhlops*
  Distribution: Bonghari and Casamance, Senegal, Kouroussa and
      Youkounkoun, Guinea

Chapter 14

FAMILY TYPHLOPIDAE

Order: Serpentes
Suborder: Scolecophidia
Family: Typhlopidae

Genus: RAMPHOTYPHLOPS Fitzinger 1843
Species typica: *multilineatus* Schlegel

Ramphotyphlops braminus (Daudin 1803)   *Eryx*
    Distribution: Zanzibar, coastal Tanzania, Mozambique, Somalia, Benin,
                Togo, Ivory Coast, Cameroon, Madagascar, Comoroes,
                Mascarenes and Seychelles
    Reference: Guibe (1958); McDowell (1974)
    Typhlops microcephalus Werner: Guibe (1958)

Genus: RHINOTYPHLOPS Fitzinger 1843
Species typica: *lalandei* Schlegel

Rhinotyphlops anomalus (Bocage 1873)   *Onychocephalus*
    Distribution: Southern Angola and northern Namibia

Rhinotyphlops caecus (Dumeril 1856)   *Onychocephalus*
    Distribution: Gabon, Sierra Leone, Togo, Benin, Ghana and Western
             Zaire
    Reference:
    Typhlops avakubae Schmidt: Schmidt (1923)

Rhinotyphlops crossii (Boulenger 1893)   *Typhlops*
    Distribution: Agberi and Asaba, Nigeria

Rhinotyphlops feae (Boulenger 1906)   *Typhlops*
    Distribution: Sao Tome and Principe Islands, Gulf of Guinea

Rhinotyphlops gracilis (Sternfeld 1910)   *Typhlops*
    Distribution: Zambia, southwestern Tanzania and Zaire

Rhinotyphlops graueri (Sternfeld 1912)   *Typhlops*
    Distribution: Zaire, Rwanda, Burundi and Ujuji, Tanzania
    Reference:Laurent (1956a)

Rhinotyphlops kibarae (Witte 1953)   *Typhlops*
    Distribution: Katanga Province, Zaire

Rhinotyphlops lalandei (Schlegel 1839)   *Typhlops*
    Distribution: South Africa and southwestern Zimbabwe
    Reference: Fitzsimons (1974)

Rhinotyphlops leucocephalus (Parker 1930)   *Typhlops*
    Distribution: Las Anod, Somalia

Rhinotyphlops lumbriciformis (Peters 1874)   *Onchocephalus*
    Distribution: Coastal Kenya and Tanzania

Rhinotyphlops newtoni (Bocage 1890)   *Typhlops*
    Distribution: Rolas, Sao Tome and Principe Islands, Gulf of Guinea

Rhinotyphlops pallidus (Cope 1868)   *Letheobia*
    Distribution: Zanzibar, Pemba Island, coastal Kenya and Tanzania;
            southern Sudan

Rhinotyphlops praeocularis (Stejneger 1894)   *Typhlops*
    Distribution: Zaire

Rhinotyphlops rufescens (Chabanaud 1916)   *Typhlops*
    Distribution: Haut-Oubangui, Central African Republic and Bambesa-
            Takpwo, lower Uele, Zaire

Rhinotyphlops schinzi (Boettger 1887)   *Typhlops*
    Distribution: Northwestern Cape Province and southern Namibia
    Reference: Fitzsimons (1974)

Rhinotyphlops schlegelii schlegelii (Bianconi 1850)   *Typhlops*
   Distribution: Southern Mozambique, northern Zululand, Swaziland to
          eastern and northern Transvaal south of the Limpopo River
   Reference: Fitzsimons (1974)

Rhinotyphlops schlegelii brevis (Scortecci 1929)   *Typhlops*
   Distribution: Southern Sudan, Somalia, central Ethiopia and northern
          Uganda

Rhinotyphlops schlegelii mucruso (Peters 1854)   *Onychocephalus*
   Distribution North of the Zambesi in the east through Mozambique
          and Tanzania to Kenya and west through Malawi and
          Zambia to Angola, northern Namibia and Botswana
   Reference: Fitzsimons (1974)

Rhinotyphlops schlegelii petersii (Bocage 1873)   *Onychocephalus*
   Distribution: Southwestern Angola

Rhinotyphlops scortecci (Gans and Laurent 1965)   *Typhlops*
   Distribution: Benadir Province, Somalia

Rhinotyphlops somalicus (Boulenger 1895)   *Typhlops*
   Distribution: Highlands of Ethiopia and western Somalia

Rhinotyphlops stejnegeri (Loveridge 1931)   *Typhlops*
   Distribution: Kasai Province, Zaire

Rhinotyphlops sudanensis (Schmidt 1923)   *Typhlops*
   Distribution: Orientale Province, Zaire

Rhinotyphlops unitaeniatus (Peters 1878)   *Typhlops*
   Distribution: Somalia, southeastern Ethiopia, Kenya and northeastern
          Tanzania

Rhinotyphlops wittei Roux-Esteve 1974
   Distribution: Province Equateur, Zaire

Genus: TYPHLOPS Oppel 1811
Species typica: *lumbricalis* Linnaeus
Reference: Roux-Esteve (1974)

Typhlops angolensis (Bocage 1866)   *Onychocephalus*
    Distribution: Zaire, Angola, coastal Cameroon, southern Central
            African Republic, Gabon, Congo, western Kenya, Uganda,
            northern Tanzania and northern Zambia
    Reference: Laurent (1964c); Roux-Esteve (1970)

Typhlops arenarius (Grandidier 1872)   *Onychocephalus*
    Distribution: Madagascar
    Reference: Guibe (1958)

Typhlops bibronii (Smith 1846)   *Onychocephalus*
    Distribution: Eastern Cape Province to Natal, Swaziland and Transvaal
    Reference: Fitzsimons (1974)

Typhlops boylei Fitzsimons 1932
    Distribution: Namibia
    Reference: Fitzsimons (1974)

Typhlops coecatus Jan 1864
    Distribution: Ghana and Ivory Coast

Typhlops comoroensis Boulenger 1889
    Distribution: Comoro Islands
    Reference: Guibe (1958)

Typhlops cuneirostris Peters 1879
    Distribution: Somalia

Typhlops decorosus Buchholz and Peters 1875
    Distribution: Cameroon

Typhlops decorsei Mocquard 1901
    Distribution: Madagascar
    Reference: Guibe (1958)

Typhlops elegans Peters 1868
    Distribution: Principe Island, Gulf of Guinea

Typhlops fornasinii Bianconi 1847
    Distribution: Coastal Mozambique, Zululand, Natal and southeastern
            Zimbabwe
    Reference:
    Typhlops mossambicus (Peters): Fitzsimons (1974)
    Typhlops tettensis tettensis (Peters): Isemonger (1962)

Typhlops gierrai Mocquard 1897
Distribution: Usambara and Uluguru Mountains, Tanzania
Reference: Laurent (1964c)

Typhlops grandidieri Mocquard 1905
Distribution: Madagascar
Reference: Guibe (1958)

Typhlops leucostictus Boulenger 1898
Distribution: Liberia

Typhlops lineolatus lineolatus Jan 1863
Distribution: Senegal, Gambia, Liberia, Guinea, Sierra Leone, Ivory
Coast, Ghana, Nigeria, Cameroon, southwest Chad,
southern Sudan, Ethiopia, Uganda, southwest Kenya,
north and central Tanzania, western Zambia, Zaire,
Angola and Congo

Typhlops lineolatus tanganicanus Laurent 1964
Distribution: Southeastern Tanzania

Typhlops madagascariensis Boettger 1887
Distribution: Madagascar
Reference: Guibe (1958)

Typhops manni Loveridge 1941
Distribution: Liberia and Guinea

Typhlops mucronatus Boettger 1880
Distribution: Madagascar
Reference: Guibe (1958)

Typhlops obtusus Peters 1865
Distribution: Malawi, Mozambique and Zimbabwe
Reference: Loveridge (1953)
Typhlops obtusus palgravei Laurent: Fitzsimons (1974)

Typhlops ocularis Parker 1927
Distribution: Madagascar
Reference: Guibe (1958)

Typhlops platyrhynchus Sternfeld 1910
Distribution: Tanga, Tanzania

Typhlops punctatus punctatus (Leach 1819)   *Acontias*
    Distribution: Senegal, Gambia, Guinea Bissau, central Mali, Ivory
                 Coast, Ghana, Togo, Benin, Nigeria, Niger, southern
                 Chad, Central African Republic, Cameroon, Uganda and
                 southwest Ethiopia
    Reference: Rieppel (1979)

Typhlops punctatus congestus (Dumeril and Bibron 1844)
                                                        Onychocephalus
    Distribution: Nigeria, Cameroon, Fernando Po, Central African
                 Republic, Gabon, Zaire, Congo and Uganda

Typhlops punctatus liberiensis (Hallowell 1848)   *Onychocephalus*
    Distribution: Sierra Leone, Liberia, Guinea, Ivory Coast and Ghana

Typhlops reuteri Boettger 1881
    Distribution: Madagascar
    Reference: Guibe (1958)

Typhlops rondoensis Loveridge 1942
    Distribution: Rondo Plateau and Newala region, Tanzania

Typhlops socotranus Boulenger 1889
    Distribution: Socotra

Typhlops steinhausi Werner 1909
    Distribution: Coastal Nigeria, southern Cameroon and southeastern
                 Central African Republic

Typhlops uluguruensis Barbour and Loveridge 1928
    Distribution: Uluguru Mountains, Tanzania

Typhlops vermicularis Merrem 1820
    Distribution: Egypt
    Reference: Arnold and Burton (1978); Heyder (1968a) (1968b) (1973);
                 Rieppel (1979)

Typhlops verticalis (Smith 1846)   *Onychocephalus*
    Distribution: Western Cape Province
    Reference: Fitzsimons (1974)

Typhlops zenkeri Sternfeld 1908
    Distribution: Cameroon

Chapter 15

FAMILY BOIDAE

Order: Serpentes
Suborder: Alethinophidia
Infraorder: Henophidia
Superfamily: Booidea
Family: Boidae
Subfamily: Pythoninae
Tribe: Calabarini

Genus: CALABARIA Gray 1858
Species typica: *reinhardtii* Schlegel

Calabaria reinhardtii (Schlegel 1848)   *Eryx*
    Distribution: Sierra Leone east to northern Zaire
    Reference: Gabe and Saint Girons (1969); Gartlan and Struhsaker
            (1971); Green (1973); Haas (1930) (1931a); Langebartel (1968)

Subfamily: Pythoninae
Tribe: Pythonini

Genus: PYTHON Daudin 1803
Species typica: *molurus*

Python anchietae Bocage 1887
    Distribution: Southern Angola and northern Namibia
    Reference: Finkeldey (1963); Fitzsimons (1974); Laurent (1964b);
            Patterson (1978); Steyn and Els (1963)

Python regius (Shaw 1802)   *Boa*
    Distribution: Senegal east to Uganda
    Reference: Edmondson (1976); Hawes (1974); Phisalix (1914) (1922);
            Schmidt and Inger (1957); Vogel (1964)

Python sebae (Gmelin 1788)   *Coluber*
    Distribution: Senegal east to Sudan and Uganda south to the Cape
    Reference: Aruo (1977); Broadley (1961b); Branch and Patterson (1975);
            Fitzsimons (1974); Frazzetta (1966); Haas (1931a);
            Jacobshagen (1920); Langebartel (1968); Loveridge (1929)
            (1931a); Patterson (1974); Schmidt and Inger (1957);
            Stemmler (1971); Vogel (1964)

Subfamily: Boinae
Tribe: Boini

Genus: ACRANTOPHIS Jan 1860
Species typica: *dumerilii* Jan

Acrantophis dumerilii Jan 1860
    Distribution: Madagascar and the Mascarenes

Acrantophis madagascariensis (Dumeril and Bibron 1844)   *Pelophilus*
    Distribution: Madagascar
    Reference: Branch and Erasmus (1976)

Genus: SANZINIA Gray 1849
Species typica: *madagascariensis* Dumeril and Bibron

Sanzinia madagascariensis (Dumeril and Bibron 1844)   *Xiphosoma*
    Distribution: Madagascar
    Reference: Branch and Erasmus (1976); Carpenter, Murphy and
            Mitchell (1978)

Subfamily: Boinae
Tribe: Erycini

Genus: ERYX Daudin 1803
Species typica: *trucicus*
Reference: Matz (1974)

Eryx colubrinus (Linnaeus 1758)   *Anguis*
    Distribution: Egypt south to Kenya and west to Niger
    Reference: Langebartel (1968)
    Eryx colubrinus loveridgei Stull: Loveridge (1957)

Eryx jaculus jaculus (Linnaeus 1758)   *Anguis*
    Distribution: Saharan Morocco east to Egypt
    Reference: Haas (1930) (1931); Zavattari (1909)

Eryx muelleri (Boulenger 1892)   *Gongylophis*
    Distribution: Mauritania, Senegal and Sierra Leone east to Sudan

Eryx somalicus Scortecci 1939
    Distribution: Somalia

Chapter 16

FAMILY BOLYERIIDAE

Order: Serpentes
Suborder: Alethinophidia
Infraorder: Henophidia
Superfamily: Bolyerioidea
Family: Bolyeriidae

Genus: BOLYERIA Gray 1842
Species typica: *multicarinata* Boie

Bolyeria multicarinata (Boie 1827)   *Eryx*
    Distribution: Round Island
    Reference: Anthony and Guibe (1951b); Guibe (1958)

Genus: CASAREA Gray 1942
Species typica: *dussumieri* Schlegel

Casarea dussumieri (Schlegel 1837)   *Boa*
    Distribution: Round Island
    Reference: Anthony and Guibe (1951a) (1951b); Gabe and Saint
        Girons (1972); Frazetta (1971)

Chapter 17

FAMILY COLUBRIDAE

Order: Serpentes
Suborder: Alethinophidia
Superfamily: Colubroidea
Family: Colubridae
Comment: In the following family I have leaned heavily on the work of
Bourgeois (1968) though all readers should consult Underwood (1967);
see also Smith, Smith and Sawin (1977). Two points should be noted:
firstly the subfamilies are listed in no specific order other than alpha-
betical and secondly, those genera whose subfamily status I am not
certain of are listed under the heading 'Subfamily uncertain', a much
preferred alternative to any unnatural lumping.

Subfamily: Aparallactinae

Genus: AMBLYODIPSAS Peters 1849
Species typica: *microphthalma* Bianconi
Reference: Bourgeois (1968); Broadley (1971d); Kochva and Wollberg
(1970); Witte and Laurent (1947)

Ambiyodipsas concolor (Smith 1849)   *Choristodon*
      Distribution: Eastern Transvaal south to Durban
      Reference: Fitzsimons (1974)

Amblyodipsas dimidiata (Gunther 1888)   *Rhinocalamus*
      Distribution: Northern Tanzania

137

Amblyodipsas katangensis katangensis Witte and Laurent 1942
   Distribution: Eastern Zaire and northern Zambia

Amblyodipsas katangensis ionidesi Loveridge 1951
   Distribution: Southern Tanzania

Amblyodipsas microphthalma (Bianconi 1852)   *Calamaria*
   Distribution: Southern Mozambique, northeastern Transvaal and
             northern Zululand
   Reference: Fitzsimons (1974)

Amblyodipsas polylepis polylepis (Bocage 1873)   *Calamelaps*
   Distribution: Angola, Namibia, Botswana, Zaire, Zambia, Mozambique,
             Zimbabwe, Malawi, Transvaal and Natal
   Reference: Fitzsimons (1974); Sweeny (1971)
   Calamelaps unicolor miolepis Gunther: Broadley (1959c); Witte and
             Laurent (1947)
   Calamelaps unicolor warreni Boulenger: Sweeny (1971)

Amblyodipsas polylepis hildebrandtii (Peters 1877)   *Atractaspis*
   Distribution: Coastal Kenya and Tanzania

Amblyodipsas rodhaini (Witte 1930)   *Rhinocalamus*
   Distribution: Zaire

Amblyodipsas teitana Broadley 1971
   Distribution: Mount Mbololo, Teita Hills near Voi, Kenya

Amblyodipsas unicolor (Reinhardt 1843)   *Calamaria*
   Distribution: Senegal east to southern Sudan and western Kenya
   Reference: Pitman (1974)

Amblyodipsas ventrimaculata (Roux 1907)   *Rhinocalamus*
   Distribution: Northern Namibia and Botswana, northwest Zimbabwe
             and western Zambia
   Reference: Fitzsimons (1974)

Genus: **APARALLACTUS** Smith 1849
Species typica: *capensis* Smith
Reference: Bourgeois (1968); Witte and Laurent (1947)

Aparallactus capensis capensis Smith 1849
Distribution: South Africa, Zimbabwe, Mozambique, Malawi, Tanzania
southern Kenya
Reference: Fitzsimons (1974); Sweeny (1971)

Aparallactus capensis bocagei Boulenger 1895
Distribution: Angola, southeastern Zaire and northern Namibia
Reference: Fitzsimons (1974)

Aparallactus capensis luebberti Sternfeld 1910
Distribution: Namibia

Aparallactus capensis nigriceps (Peters 1854)   *Uriechis*
Distribution: Southern Mozambique

Aparallactus capensis punctatolineatus Boulenger 1895
Distribution: Eastern Angola, Zaire, western Tanzania and northern
Malawi

Aparallactus capensis uluguruensis Barbour and Loveridge 1928
Distribution: Border areas of Kenya and Tanzania

Aparallactus guentheri Boulenger 1895
Distribution: Angola, Zambia, Malawi, northern Mozambique and
southern Tanzania
Reference: Fitzsimons (1974); Sweeny (1971)

Aparallactus jacksonii (Gunther 1888)   *Uriechis*
Distribution: Uganda, northern Tanzania, Kenya, southern Sudan and
Ethiopia

Aparallactus lineatus (Peters 1870)   *Uriechis*
Distribution: Liberia and Guinea east to Nigeria

Aparallactus lunulatus lunulatus (Peters 1854)   *Uriechis*
Distribution: Mozambique, Transvaal, Malawi, Zimbabwe, Tanzania,
Uganda, Kenya, Ethiopia, southern Sudan, Central African
Republic, northern Zaire, Congo, Cameroon, Nigeria,
Benin and Togo
Reference: Fitzsimons (1974); Pitman (1974); Sweeny (1971)

Aparallactus lunulatus scortecci Parker 1949
Distribution: Somalia

Aparallactus modestus modestus (Gunther 1859)   *Elapops*
    Distribution: Liberia and Sierra Leone east to Zaire

Aparallactus modestus ubangensis Boulenger 1897
    Distribution: Zaire and Uganda
    Reference: Pitman (1974)

Aparallactus moeruensis Witte and Laurent 1943
    Distribution: Zaire

Aparallactus niger Boulenger 1897
    Distribution: Sierra Leone, Guinea and Ivory Coast

Aparallactus turneri Loveridge 1935
    Distribution: Kenya

Aparallactus werneri Boulenger 1895
    Distribution: Tanzania

Genus: BRACHYOPHIS Mocquard 1888
Species typica: *revoili* Mocquard
Reference: Parker (1949); Witte and Laurent (1947)

Brachyophis revoili revoili Mocquard 1888
    Distribution: Kenya and southern Somalia

Brachyophis revoili cornii Scortecci 1932
    Distribution: Central Somalia

Genus: CHILORHINOPHIS Werner 1907
Species typica: *butleri* Werner
Reference: Bourgeois (1968); Witte and Laurent (1947)

Chilorhinophis butleri Werner 1907
    Distribution: Sudan south to Tanzania

Chilorhinophis carpenteri (Parker 1927)   *Parkerophis*
    Distribution: Mozambique

Chilorhinophis gerardi (Boulenger 1913)   *Apostolepis*
    Distribution: Zaire, Zambia and Zimbabwe
    Reference: Fitzsimons (1974)

Genus: **ELAPOCALAMUS** Boulenger 1911
Species typica: *gracilis* Boulenger

Elapocalamus gracilis Boulenger 1911
    Distribution: Cameroon
    Reference: Witte and Laurent (1947)

Genus: **ELAPOTINUS** Jan 1862
Species typica: *picteti* Jan

Elapotinus picteti Jan 1862
    Distribution: Unknown
    Reference: Witte and Laurent (1947)

Genus: **HOMOROSELAPS** Jan 1858
Species typica: *lacteus* Linnaeus
Reference: Kochva and Wollberg (1970); McDowell (1968)
Comment: *Homorelaps* Boulenger 1896, appears to be an unjustified
        emendation

Homoroselaps dorsalis (Smith 1849)   *Elaps*
    Distribution: Northeastern South Africa

Homoroselaps lacteus (Linnaeus 1758)   *Coluber*
    Distribution: Northern South Africa

Genus: **HYPOPTOPHIS** Boulenger 1908
Species typica: *wilsoni* Boulenger
Reference: Witte and Laurent (1947)

Hypoptophis wilsoni wilsoni Boulenger 1908
    Distribution: Southern Zaire

Hypoptophis wilsoni katangae (Muller 1911)   *Michellia*
    Distribution: Southeastern Zaire

Genus: **MACRELAPS** Boulenger 1896
Species typica: *microlepidotus* Gunther

Macrelaps microlepidotus (Gunther 1860)   *Uriechis*
    Distribution: Natal and Cape Province
    Reference: Fitzsimons (1974); Witte and Laurent (1947)

Genus: **MICRELAPS** Boettger 1880
Species typica: *muelleri* Boettger
Reference: Parker (1949); Witte and Laurent (1947)

Micrelaps bicoloratus Sternfeld 1908
    Distribution: Kenya and Tanzania

Micrelaps boettgeri Boulenger 1896
    Distribution: Somalia, Ethiopia, Kenya and Uganda
    Reference: Pitman (1974)

Micrelaps vaillanti (Mocquard 1888)   *Elaposchema*
    Distribution: Somalia

Genus: **POLEMON** Jan 1858
Species typica: *barthii* Jan
Reference: Bourgeois (1968); Broadley (1971d); Hughes and Barry (1969);
          Witte and Laurent (1947)

Polemon acanthias (Reinhardt 1860)   *Urobelus*
    Distribution: Guinea, Liberia, Ivory Coast, Ghana, Togo and Sierra
                Leone
    Reference: Cole (1967)

Polemon barthii Jan 1858
    Distribution: Guinea

Polemon bocourti Mocquard 1897
    Distribution: Cameroon, Rio Muni and Gabon

Polemon christyi (Boulenger 1903)   *Miodon*
    Distribution: Zaire, Uganda and Tanzania
    Reference: Pitman (1974)

Polemon collaris collaris (Peters 1881)   *Microsoma*
    Distribution: Angola north to Cameroon and east to Uganda
    Reference: McDowell (1968); Pitman (1974)

Polemon collaris brevior (Witte and Laurent 1947)   *Miodon*
    Distribution: Nigeria and Cameroon

Polemon collaris longior (Witte and Laurent 1947)   *Miodon*
    Distribution: Northeastern Zaire

Polemon fulvicollis fulvicollis (Mocquard 1887)   *Microsoma*
    Distribution: Gabon

Polemon fulvicollis gracilis (Witte and Laurent 1943)   *Miodon*
    Distribution: Western Zaire

Polemon fulvicollis graueri (Sternfeld 1908)   *Miodon*
    Distribution: Uganda and eastern Zaire
    Reference: Pitman (1974)

Polemon gabonensis gabonensis (Dumeril 1856)   *Elapomorphus*
    Distribution: Gabon, Rio Muni and Cameroon west to Togo

Polemon gabonensis schmidti (Witte and Laurent 1947)   *Miodon*
    Distribution: Zaire

Polemon griseiceps (Laurent 1947)   *Miodon*
    Distribution: Cameroon and Central African Republic

Polemon leopoldi (Witte 1941)   *Melanocalamus*
    Distribution: Rwanda

Polemon neuwiedi (Jan 1858)   *Microsoma*
    Distribution: Ivory Coast east to Nigeria

Polemon notatus notatus (Peters 1882)   *Microsoma*
    Distribution: Cameroon, Congo and Gabon

Polemon notatus aemulans (Werner 1902)   *Cynodontophis*
    Distribution: Congo and Zaire

Polemon robutus (Witte and Laurent 1943)   *Miodon*
    Distribution: Zaire

Genus: XENOCALAMUS Gunther 1868
Species typica: *bicolor* Gunther
Reference: Bourgeois (1968); Broadley (1971d); Witte and Laurent (1947)

Xenocalamus bicolor bicolor Gunther 1868
   Distribution: Namibia, western and central Botswana, northern Cape
             Province, western Orange Free State, northern Zimbabwe
             and central Mozambique
   Reference: Fitzsimons (1974)
Xenocalamus bicolor concavorostralis Hoffman: Witte and Laurent
                                       (1947)
Xenocalamus bicolor maculatus Fitzsimons: Witte and Laurent (1947)
Xenocalamus bicolor pernasutus (Werner); Witte and Laurent (1947)

Xenocalamus bicolor australis Fitzsimons 1946
   Distribution: Waterberg District, Transvaal
   Reference: Fitzsimons (1974)

Xenocalamus bicolor lineatus Roux 1907
   Distribution: Southern Mozambique, northeast Transvaal and southeast
             Zimbabwe
   Reference: Fitzsimons (1974)

Xenocalamus bicolor machadoi Laurent 1954
   Distribution: Angola and western Zaire

Xenocalamus mechowii mechowii Peters 1881
   Distribution: Northern Angola and southwestern Zaire

Xenocalamus mechowii inornatus Witte and Laurent 1947
   Distribution: Northern Namibia, Botswana, western Zimbabwe and
             Zambia
   Reference: Fitzsimons (1974)

Xenocalamus michelli Muller 1911
   Distribution: Zaire

Xenocalamus sabiensis Broadley 1971
   Distribution: Southeastern Zimbabwe
   Reference: Fitzsimons (1974)

Xenocalamus transvaalensis Methuen 1919
   Distribution: Northern Transvaal south through Mozambique to
             Zululand
   Reference: Fitzsimons (1974)

Subfamily Atractaspidinae

Genus: **ATRACTASPIS** Smith 1849
Species typica: *bibroni* Smith
Reference: Bourgeois (1968); Harding and Welch (1980); Laurent (1950b)

Atractaspis aterrima Gunther 1863
  Distribution: Guinea, Guinea Bissau, Ivory Coast, Ghana, Gabon, Zaire,
        Rwanda, Burundi, Uganda and Tanzania
  Reference: Loveridge (1957); Pitman (1974)

Atractaspis battersbyi Witte 1959
  Distribution: Zaire

Atractaspis bibroni Smith 1849
  Distribution: South Africa north to Angola and Kenya
  Reference: Broadley (1974d)
  Atractaspis bibroni rostrata Gunther: Loveridge (1957)
  Atractaspis duerdeni Gough: Laurent (1950b)

Atractaspis boulengeri boulengeri Mocquard 1897
  Distribution: Gabon

Atractaspis boulengeri matschiensis Werner 1897
  Distribution: Cameroon

Atractaspis boulengeri mixta Laurent 1945
  Distribution: Zaire

Atractaspis boulengeri schmidti Laurent 1945
  Distribution: Zaire

Atractaspis boulengeri schultzei Sternfeld 1917
  Distribution: Congo

Atractaspis boulengeri vanderborghti Laurent 1956
  Distribution: Zaire

Atractaspis coalescens Perret 1960
  Distribution: Cameroon

Atractaspis congica congica Peters 1877
 Distribution: Angola, Congo and Cameroon

Atractaspis congica leleupi Laurent 1950
 Distribution: Zaire

Atractaspis congica orientalis Laurent 1945
 Distribution: Zaire

Atractaspis corpulenta corpulenta (Hallowell 1854)   *Brachycranion*
 Distribution: Cameroon, Congo, Gabon and Zaire
 Reference: Haas (1931a) (1931b)

Atractaspis corpulenta kivuensis Laurent 1958
 Distribution: Zaire

Atractaspis corpulenta leucura Mocquard 1885
 Distribution: Ivory Coast and Ghana

Atractaspis dahomeyensis Bocage 1887
 Distribution: Guinea east to Cameroon
 Reference: Warrell, Ormerod and Davidson (1976)

Atractaspis engdahli Lonnberg and Andersson 1913
 Distribution: Somalia

Atractaspis irregularis irregularis (Reinhardt 1843)   *Elaps*
 Distribution: Guinea and Liberia east to Cameroon and south to
        northern Angola
 Reference: Britt (1978)
 Atractaspis irregularis parkeri Laurent: Harding and Welch (1980);
        Laurent (1950b)

Atractaspis irregularis angeli Laurent 1950
 Distribution: Mount Bizen, Ethiopia

Atractaspis irregularis bipostocularis Boulenger 1905
 Distribution: Kenya, northeastern Tanzania, Uganda, Rwanda and
        northeastern Zaire
 Reference: Loveridge (1957); Pitman (1974)
 Atractaspis irregularis conradsi Sternfeld: Bourgeois (1963); Laurent
        (1950b)
 Atractaspis irregularis loveridgei Laurent: Laurent (1950b)

Atractaspis irregularis uelensis Laurent 1945
   Distribution: Northeastern Zaire, northwestern Uganda and southern
            Sudan
   Reference: Lanza and Vanni (1976); Pitman (1974)

Atractaspis leucomelas Boulenger 1895
   Distribution: Ethiopia and Somalia

Atractaspis microlepidota microlepidota Gunther 1866
   Distribution: Sudan, Ethiopia, Somalia and Kenya
   Reference: Corkill and Kirk (1954); Kochva (1962); Langebartel (1968);
            Loveridge (1957); Minton (1968); Pitman (1974); Warrell,
            Ormerod and Davidson (1976)
   Atractaspis microlepidota fallax Peters: Laurent (1950b)

Atractaspis microlepidota magrettii Scortecci 1928
   Distribution: Ethiopia and southeastern Sudan

Atractaspis microlepidota micropholis Gunther 1872
   Distribution: Mauritania, northern Nigeria, Mali and Niger

Atractaspis reticulata reticulata Sjostedt 1896
   Distribution: Cameroon

Atractaspis reticulata brieni Laurent 1956
   Distribution: Zaire

Atractaspis reticulata heterochilus Boulenger 1901
   Distribution: Cameroon, Rio Muni, Gabon and Congo

Atractaspis scorteccii Parker 1949
   Distribution: Somalia

Subfamily Boiginae

Genus: CROTAPHOPELTIS Fitzinger 1843
Species typica: *hotamboeia* Laurenti
Reference: Bourgeois (1968)

Crotaphopeltis acarina Roman 1974
   Distribution: Upper Volta

Crotaphopeltis barotseensis Broadley 1968
Distribution: Zambia

Crotaphopeltis degeni (Boulenger 1906)   *Leptodira*
Distribution: Southern Sudan, Uganda and Kenya
Reference: Loveridge (1957); Pitman (1974)

Crotaphopeltis hotamboeia (Laurenti 1768)   *Coronella*
Distribution: Africa south of 15 °N, excluding Somalia
Reference: Brock (1929); Fitzsimons (1974); Loveridge (1957); Pitman
        (1974)
    Crotaphopeltis hotamboeia bicolor Leach: Bourgeois (1968); Witte
                (1962)
    Crotaphopeltis hotamboeia kageleri Uthmoller: Loveridge (1957)
    Crotaphopeltis hotamboeia ruziziensis Laurent: Witte (1962)
    Crotaphopeltis hotamboeia tornieri (Werner): Loveridge (1957)

Genus: DIPSADOBOA Gunther 1858
Species typica: *unicolor* Gunther
Reference: Bourgeois (1968)

Dipsadoboa aulicus aulicus (Gunther 1864)   *Chamaetortus*
Distribution: Southeastern Kenya, Tanzania, Mozambique, southeastern
            Zimbabwe, eastern Transvaal and Zululand

Dipsadoboa aulicus flavidus (Broadley and Stevens 1971)   *Chamaetortus*
Distribution: Malawi

Dipsadoboa duchesnii (Boulenger 1901)   *Leptodira*
Distribution: Zaire west to Sierra Leone
Reference: Laurent (1956a)

Dipsadoboa elongata elongata (Barbour 1914)   *Crotaphopeltis*
Distribution: Cameroon south to Zaire

Dipsadoboa elongata gracilis Laurent 1956
Distribution: Zaire

Dipsadoboa shrevei (Loveridge 1932)   *Crotaphopeltis*
Distribution: Zaire, Angola and Zambia
Reference: Laurent (1956a) (1964b)

Dipsadoboa unicolor unicolor Gunther 1858
  Distribution: Uganda, northern Zaire and Central African Republic
      west to Guinea

Dipsadoboa unicolor viridiventris Laurent 1956
  Distribution: Uganda, Rwanda, Burundi and adjacent Zaire

Dipsadoboa werneri (Boulenger 1897)   *Leptodira*
  Distribution: Usambara Mountains, Tanzania

Genus: MALPOLON Fitzinger 1826
Species typica: *monspessulanus* Hermann

Malpolon miolensis (Reuss 1834)   *Coluber*
  Distribution: Saharan Morocco east to Egypt
  Reference: Domergue (1959b); Schmidt and Inger (1957)

Malpolon monspessulanus monspessulanus (Hermann 1804)   *Coluber*
  Distribution: West and north Morocco
  Reference: Arnold and Burton (1978); Dunson, Dunson and Keith
      (1978); Minton and Salanitro (1972); Schmidt and Inger
      (1957); Steward (1971)

Malpolon monspessulanus insignitus (Geoffroy 1827)   *Coluber*
  Distribution: Extreme eastern Morocco east to Egypt
  Reference: Steward (1971)

Genus: TELESCOPUS Wagler 1830
Species typica: Coluber on pl.v in Geoffroy (1812)
Reference: Bourgeois (1968)

Telescopus beetzi (Barbour 1922)   *Tarbophis*
  Distribution: Namibia and northwestern Cape Province
  Reference: Fitzsimons (1974)

Telescopus dhara obtusus (Reuss 1834)   *Coluber*
  Distribution: Egypt south to Somalia and Uganda west to Mauritania
  Reference: Isemonger (1962); Marx (1968)

Telescopus dhara somalicus (Parker 1949)   *Tarbophis*
  Distribution: Southern Somalia and adjacent Kenya
  Reference: Isemonger (1962)

Telescopus guidimakaensis (Chabanaud 1916)   *Tarbophis*
    Distribution: Mauritania

Telescopus pulcher (Scortecci 1935)   *Migiurtinophis*
    Distribution: Somalia
    Reference: Parker (1949)

Telescopus semiannulatus semiannulatus Smith 1849
    Distribution: Kenya, Tanzania and Congo south to Namibia, Botswana
            and Swaziland
    Reference: Fitzsimons (1974); Saiff (1975)

Telescopus semiannulatus polystictus Mertens 1954
    Distribution: Highlands of southern Damaraland, Namibia
    Reference: Fitzsimons (1974)

Telescopus variegatus (Reinhardt 1843)   *Tarbophis*
    Distribution: Sierra Leone east to Niger and Cameroon

Genus: TOXICODRYAS Hallowell 1857
Species typica: *blandingii* Hallowell
Reference: Bourgeois (1968); Laurent (1956a) (1964b); Loveridge (1957)
Comment: Underwood (1967) includes the underlisted species in the family
    Dipsadidae: Lycodontinae while placing the Asiatic *Boiga* in the family
    Homalopsidae: Boiginae. To reflect such a situation I herein use the next
    available generic name *Toxicodryas* for the African species whose af-
    finities probably lay with the *Dipsadoboa-Crotaphopeltis-Telescopus*
    group, a group which is in need of full investigation.

Toxicodryas blandingii (Hallowell 1844)   *Dipsas*
    Distribution: Western Kenya, Uganda, southern Sudan, Zaire, Angola,
            Gabon, Congo, Cameroon, Equatorial, Guinea, Central
            African Republic, Nigeria, Benin, Togo, Ghana, Ivory
            Coast, Liberia, Sierra Leone and Guinea
    Reference: Groves (1973); Levinson, Evans and Groves (1976)

Toxicordryas pulverulenta (Fischer 1856)   *Dipsas*
    Distribution: Liberia, Sierra Leone and Guinea east to Uganda, Zaire
            and Angola

Subfamily Boodontinae

Genus: BOTHROLYCUS Gunther 1874
Species typica: *ater* Gunther

Bothrolycus ater Gunther 1874
   Distribution: Cameroon south to Zaire
   Reference: Dowling (1969); Laurent (1956a)

Genus: BOTHROPHTHALMUS Peters 1863
Species typica: *lineatus* Peters

Bothrophthalmus lineatus lineatus (Peters 1863)   *Elaphis*
   Distribution: Angola, Zaire and Uganda west to Sierra Leone
   Reference: Bourgeois (1968); Dowling (1969); Laurent (1964b)

Bothrophthalmus lineatus brunneus Gunther 1863
   Disribution: Northern Gabon, Equatorial Guinea and Cameroon

Genus: LAMPROPHIS Fitzinger 1843
Species typica: *aurora* Linnaeus
Reference: Bourgeois (1968); Dowling (1969); Fitzsimons (1962) (1974);
         Roux-Esteve and Guibe (1965a) (1965b); Thorpe and McCarthy
         (1978)
Comment: Herein I include the genus *Boaedon* of other authors.

Lamprophis aurora (Linnaeus 1758)   *Coluber*
   Distribution: Cape Province north to Transvaal and east to Swaziland

Lamprophis erlangeri (Werner 1923)   *Pseudoboodon*
   Distribution: Ethiopia

Lamprophis fiskii Boulenger 1887
   Distribution: Western Cape Province

Lamprophis fuliginosus (Boie 1827)   *Lycodon*
   Distribution: Africa south of the Sahara west and north to Morocco and
             Mauritania
   Reference: Gabe and Saint Girons (1969); Loveridge (1957); Pasteur
             and Bons (1960)

Boaedon fuliginosus bedriagae Boulenger: Roux-Esteve and Guibe
(1965b)
Boaedon fuliginosus mentalis Gunther: Fitzsimons (1974); Roux-
Esteve and Guibe (1965b)

Lamprophis fuscus Boulenger 1893
Distribution: Cape Province east to Natal and southeastern Mozambique

Lamprophis geometricus (Schlegel 1837)   *Lycodon*
Distribution: Seychelles

Lamprophis guttatus (Smith 1843)   *Lycodon*
Distribution: Cape Province north to central Namibia and eastern
Transvaal
Reference: Haacke (1965)

Lamprophis inornatus Dumeril and Bibron 1854
Distribution: Western Cape Province east to Natal and then north to
Transvaal
Reference: Pringle (1954)

Lamprophis maculatus (Parker 1932)   *Boaedon*
Distribution: Somalia
Reference: Lanza (1978c); Parker (1949)

Lamprophis olivaceus (Dumeril 1856)   *Holuropholis*
Distribution: Uganda, Zaire and Angola west to Guinea and Liberia

Lamprophis swazicus Schaefer 1970
Distribution: Swaziland

Lamprophis virgatus (Hallowell 1854)   *Coelopeltis*
Distribution: Guinea east to Zaire; the islands of the Gulf of Guinea

Genus: LYCODONOMORPHUS Fitzinger 1843
Species typica: *rufulus* Lichtenstein
Reference: Bourgeois (1968); Broadley (1967a); Fitzsimons (1962) (1974);
Loveridge (1958)

Lycodonomorphus bicolor (Gunther 1894)   *Glypholycus*
Distribution: Shores of Lake Tanganyika

Lycodonomorphus laevissimus laevissimus (Gunther 1862)   *Natrix*
    Distribution: Eastern Cape Province

Lycodonomorphus laevissimus fitzsimonsi Raw 1973
    Distribution: Natal north into southeastern Transvaal

Lycodonomorphus laevissimus natalensis Raw 1973
    Distribution: Natal

Lycodonomorphus leleupi leleupi (Laurent 1950)   *Ablabophis*
    Distribution: Zaire

Lycodonomorphus leleupi mlanjensis Loveridge 1953
    Distribution: Southern Malawi and eastern Zimbabwe

Lycodonomorphus rufulus (Lichtenstein 1823)   *Coluber*
    Distribution: Zimbabwe and southern Mozambique south to Cape
        Province and Natal

Lycodonomorphus subtaeniatus subtaeniatus Laurent 1954
    Distribution: Northern Angola and western Zaire

Lycodonomorphus subtaeniatus upembae Laurent 1954
    Distribution: Southeastern Zaire

Lycodonomorphus whytii (Boulenger 1897)   *Glypholycus*
    Distribution: Tanzania south to Transvaal

Genus: PSEUDOBOODON Peracca 1897
Species typica: *lemniscatum* Dumeril and Bibron

Pseudoboodon lemniscatus (Dumeril and Bibron 1854)   *Boaedon*
    Distribution: Ethiopia
    Reference: Dowling (1969); Parker (1949)

Subfamily Colubrinae

Genus: AELUROGLENA Boulenger 1898
Species typica: *cucullata* Boulenger

Aeluroglena cucullata Boulenger 1898
   Distribution: Somalia
   Reference: Bogert (1940); Parker (1949)

Genus: CORONELLA Laurenti 1768
Species typica: *austriaca* Laurenti

Coronella girondica (Daudin 1803)   *Coluber*
   Distribution: Morocco, Algeria and Tunisia
   Reference: Arnold and Burton (1978); Bogert (1940); Pasteur and Bons
            (1960); Steward (1971)
   Coronella amaliae (Boettger): Domergue (1959a)

Genus: HAEMORRHOIS Boie 1826
Species typica: *hippocrepis* Linnaeus
Reference: Inger and Clark (1943); Parker (1949); Welch (1980)

Haemorrhois algirus algirus (Jan 1863)   *Zamenis*
   Distribution: Libya west to Morocco and Mauritania
   Reference: Arnold and Burton (1978); Bons, J. (1962); Vogel (1964)

Haemorrhois algirus villiersi (Bons, J. 1962)   *Coluber*
   Distribution: Morocco

Haemorrhois brevis brevis (Boulenger 1895)   *Zamenis*
   Distribution: Somalia

Haemorrhois brevis boschisi (Scortecci 1930)   *Zamenis*
   Distribution: Somalia

Haemorrhois citernii (Boulenger 1912)   *Zamenis*
   Distribution: Southern Somalia and northern Kenya

Haemorrhois florulentus florulentus (Geoffroy 1827)   *Coluber*
   Distribution: Libya, Egypt and Sudan
   Reference: Bons, J. (1962)

Haemorrhois florulentus smithi (Boulenger 1895)   *Zamenis*
   Distribution: Ethiopia, Somalia, Kenya and Uganda

Haemorrhois hippocrepis hippocrepis (Linnaeus 1758)   *Coluber*
    Distribution: Morocco, northern Algeria and Tunisia
    Reference: Arnold and Burton (1978); Bons, J. (1962); Bruno and Hotz
        (1976); Pasteur and Bons (1960); Steward (1971)

Haemorrhois hippocrepis intermedius (Werner 1929)   *Coluber*
    Distribution: Saharan Morocco and Algeria
    Reference: Steward (1971)

Haemorrhois keniensis (Parker 1932)   *Coluber*
    Distribution: Lake Baringo, Kenya

Haemorrhois ravergieri ravergieri (Menetries 1832)   *Coluber*
    Distribution: Egypt
    Reference: Minton and Salanitro (1972); Vogel (1964)

Haemorrhois rhodorhachis rhodorhachis (Jan 1865)   *Zamenis*
    Distribution: Libya and Egypt
    Reference: Vogel (1964)

Haemorrhois rhodorhachis subnigra (Boettger 1893)   *Zamenis*
    Distribution: Somalia and Eritrea, Ethiopia

Haemorrhois rogersi (Anderson 1893)   *Zamenis*
    Distribution: Eastern Libya and Egypt

Haemorrhois socotrae (Gunther 1881)   *Zamenis*
    Distribution: Socotra

Haemorrhois somalicus (Boulenger 1896)   *Zamenis*
    Distribution: Somalia

Haemorrhois taylori (Parker 1949)   *Coluber*
    Distribution: Somalia

Genus: LYTORHYNCHUS Peters 1862
Species typica: *diadema* Dumeril and Bibron

Lytorhynchus diadema (Dumeril and Bibron 1854)   *Heterodon*
    Distribution: Mauritania, Morocco, Algeria, Tunisia, Libya, Egypt and
          Niger
    Reference: Flower (1933); Leviton and Anderson (1970); Pasteur and
          Bons (1960)

Genus: MEIZODON Fischer 1856
Species typica: *coronatus* Schlegel

Meizodon coronatus (Schlegel 1837)   *Calamaria*
    Distribution: Africa between latitudes 15°N and 5°S

Meizodon semiornatus (Peters 1854)   *Coronella*
    Distribution: Sudan, Ethiopia, Somalia and Uganda south to Zimbabwe,
        Transvaal and Swaziland
    Reference:
    Meizodon loveridgei Bogert: Bogert (1940)
    Meizodon somalicus (Scortecci): Bogert (1940); Scortecci (1932)

Genus: SPALEROSOPHIS Jan 1865
Species typica: *microlepis* Jan
Reference: Marx (1959)

Spalerosophis diadema cliffordi (Schlegel 1837)   *Coluber*
    Distribution: Western Sahara, Mauritania, Mali, Niger, Algeria, Tunisia,
        Libya, Egypt and northern Sudan
    Reference: Dmi'el (1967); Dmi'el and Borut (1972); Dmi'el and Zilber
        (1971); Pasteur (1967); Pasteur and Bons (1960)

Spalerosophis dolichospilus (Werner 1923)   *Zamenis*
    Distribution: Morocco, Algeria and Tunisia
    Reference: Pasteur (1967); Pasteur and Bons (1960)
    Coluber choumowitchi Domergue: Domergue (1954) (1959b)

Spalerosophis josephscorteccii Lanza 1964
    Distribution: Somalia
    Reference: Lanza (1978c)

Subfamily Dasypeltinae

Genus: DASYPELTIS Wagler 1830
Species typica: *scaber* Linnaeus
Reference: Gans (1959)

Dasypeltis atra Sternfeld 1912
    Distribution: Zaire, Uganda, Kenya and southern Sudan
    Reference: Gans and others (1968); Laurent (1956a)

Dasypeltis fasciata Smith 1849
 Distribution: Western Uganda, northern Zaire, Central Africa Republic
 and southern Chad west to Gambia
 Reference: Forcart (1964); Gans (1960a)

Dasypeltis inornata Smith 1849
 Distribution: Natal, southeastern Cape Province and Swaziland

Dasypeltis medici medici (Bianconi 1859)   *Dipsas*
 Distribution: Southern Kenya, Tanzania, north and central Mozambique
 inland to Malawi and northeastern Zimbabwe

Dasypeltis medici lamuensis Gans 1957
 Distribution: Somalia south to Kenya-Tanzania border

Dasypeltis palmarum (Leach 1818)   *Coluber*
 Distribution: Angola, Zaire and Congo

Dasypeltis scabra (Linnaeus 1758)   *Coluber*
 Distribution: Egypt to Somalia, south to the Cape and west to Gambia
 Reference: Cogger (1966); Gabe and Saint Girons (1969); Gans (1952);
 Haas (1931a) (1931b) (1932); Langebartel (1968); Laurent
 (1956a); Pringle (1954); Rabb and Snediger (1960); Schmidt
 and Inger (1957)

Subfamily Dispholidinae

Genus: DISPHOLIDUS Duvernoy 1832
Species typica: *typus* Smith
Reference: Bourgeois (1968)

Dispholidus typus (Smith)
 Reference: Broadley (1962a); Gabe and Saint Girons (1969) (1972);
 Grasset and Schaafsma (1940a) (1940b); Guillin, Bezeaud
 and Menach (1978); Haas (1931b); Hiestand and Hiestand
 (1979); Mackay and others (1969); Robertson and Delpierre
 (1969); Schaefer (1975); Schmidt and Inger (1957); Vogel
 (1964)

Dispholidus typus typus (Smith 1829)   *Bucephalus*
 Distribution: Savannah areas of Africa south of the Sahara avoiding the
 extremely dry areas

Dispholidus typus kivuensis Laurent 1955
  Distribution: Zambia, Zaire, Uganda, Rwanda and Kenya

Dispholidus typus punctatus Laurent 1955
  Distribution: Southern Zaire, Angola and Zambia

Genus: RHAMNOPHIS Gunther 1862
Species typica: *aethiopissa* Gunther
Reference: Bourgeois (1968); Loveridge (1944c)

Rhamnophis aethiopissa aethiopissa Gunther 1862
  Distribution: Senegal, Sierra Leone and Liberia east to Zaire; Fernando Po

Rhamnophis aethiopissa elgonensis Loveridge 1929
  Distribution: Western Kenya and Uganda

Rhamnophis aethiopissa ituriensis Schmidt 1923
  Distribution: Zaire

Rhamnophis batesii (Boulenger 1908)   *Thrasops*
  Distribution: Cameroon south to Zaire

Genus: THELOTORNIS Smith 1849
Species typica: *capensis* Smith
Reference: Bourgeois (1968); Loveridge (1944c)

Thelotornis capensis capensis Smith 1849
  Distribution: Tanzania, Zambia, Malawi, Mozambique, southern
            Zimbabwe, Botswana, north and east Transvaal, eastern
            Natal and Namibia
  Reference: Broadley (1957c); Fitzsimons (1974)

Thelotornis capensis oatesi (Gunther 1881)   *Dryiophis*
  Distribution: Northern Zimbabwe, Zambia, Botswana and adjacent
            Angola and Zaire

Thelotornis kirtlandii (Hallowell 1844)   *Leptophis*
  Distribution: Somalia, Kenya, Tanzania, Uganda, Zaire, Angola, Congo,
            Gabon, Rio Muni, Nigeria, Cameroon, Togo, Benin,
            Ghana, Ivory Coast, Liberia, Sierra Leone, Guinea and
            Guinea Bissau

Reference: Beiran and Currie (1967); Chapman (1968); Fitzsimons
(1962) (1974); Goodman and Goodman (1976); Kornalik,
Erbanova and Mebs (1978); Kornalik, Mebs and Taborska
(1978); Kornalik and Taborska (1978); Kornalik, Taborska
and Mebs (1978); Schaefer (1979); Sweeny (1961); Taborska,
Mebs and Kornalik (1978)

Genus: THRASOPS Hallowell 1857
Species typica: *flavigularis* Hallowell
Reference: Bourgeois (1968); Loveridge (1944c)

Thrasops flavigularis (Hallowell 1852)   *Dendrophis*
    Distribution: Sierra Leone east to Congo and Zaire

Thrasops jacksonii jacksonii Gunther 1895
    Distribution: Tanzania, Kenya, Uganda, Rwanda, Burundi and eastern
        Zaire

Thrasops jacksonii schmidti Loveridge 1936
    Distribution: Forests of Mount Kenya and Muthaiga near Nairobi, Kenya

Thrasops occidentalis Parker 1940
    Distribution: Guinea east to Togo

Subfamily Geodipsadinae

Genus: GEODIPSAS Boulenger 1896
Species typica: *infralineatus* Gunther
Reference: Bourgeois (1968)

Geodipsas boulengeri (Peracca 1892)   *Tachymensis*
    Distribution: Madagascar
    Reference: Guibe (1958)

Geodipsas depressiceps depressiceps (Werner 1897)   *Tropidonotus*
    Distribution: Cameroon, Rio Muni, Gabon, Congo, northern Zaire,
        Central African Republic and Uganda
    Reference: Dowling (1969); Laurent (1956a)

Geodipsas depressiceps marlieri Laurent 1956
    Distribution: Zaire

Geodipsas heimi Angel 1936
    Distribution: Madagascar
    Reference: Guibe (1958)

Geodipsas infralineata (Gunther 1882)   *Tachymensis*
    Distribution: Madagascar
    Reference: Guibe (1958)

Geodipsas proctera Loveridge 1922
    Distribution: Uluguru Mountains, Tanzania

Geodipsas vauerecegae Tornier 1902
    Distribution: Usambara, Uluguru and Magrotto Mountains, Tanzania

Subfamily Lycodontinae

Genus: LYCODON Fitzinger 1826
Species typica: *aulicus* Linnaeus

Lycodon aulicus aulicus (Linnaeus 1754)   *Coluber*
    Distribution: Seychelles and the Mascarenes
    Reference: Seshadri (1959)

Subfamily Lycophidinae

Genus: CHAMAELYCUS Boulenger 1919
Species typica: *christyi* Boulenger
Reference: Witte (1963)

Chamaelycus christyi Boulenger 1919
    Distribution: Zaire

Chamaelycus fasciatus (Gunther 1858)   *Alopecion*
    Distribution: Zaire and Gabon west to Sierra Leone
    Reference: Broadley (1968d); Dowling (1969); Laurent (1956a)
    Chamaelycus werneri (Mocquard): Dowling (1969); Witte (1963)

Chamaelycus parkeri (Angel 1934) *Oophilositum*
    Distribution: Zaire and Angola
    Reference: Laurent (1956a)

Genus: CRYPTOLYCUS Broadley 1968
Species typica: *nanus* Broadley

Cryptolycus nanus Broadley 1968
    Distribution: Mozambique

Genus: DENDROLYCUS Laurent 1956
Species typica: *elapoides* Schmidt

Dendrolycus elapoides elapoides (Schmidt 1923) *Lycophidion*
    Distribution: Zaire

Dendrolycus elapoides angusticinctus (Laurent 1952) *Lycophidion*
    Distribution: Zaire

Genus: GONIONOTOPHIS Boulenger 1893
Species typica: *brussauxi* Mocquard
Reference: Loveridge (1939)

Gonionotophis brussauxi brussauxi (Mocquard 1889) *Godionotus*
    Distribution: Cameroon and Congo
    Reference: Dowling (1969)

Gonionotophis brussauxi prigoginei Laurent 1956
    Distribution: Zaire

Gonionotophis grantii (Gunther 1863) *Simocephalus*
    Distribution: Guinea Bissau east to Cameroon

Gonionotophis klingi Matschie 1893
    Distribution: Togo

Genus: HELOPHIS Witte and Laurent 1942
Species typica: *schoutedeni* Witte

Helophis schoutedeni (Witte 1922) *Pelophis*
    Distribution: Zaire

Genus: HORMONOTUS Hallowell 1857
Species typica: *modestus* Dumeril and Bibron

Hormonotus modestus (Dumeril and Bibron 1854)   *Lamprophis*
    Distribution: Angola, Zaire and Uganda west to Guinea
    Reference: Bourgeois (1968); Dowling (1969)

Genus: HYDRAETHIOPS Gunther 1872
Species typica: *melanogaster* Gunther

Hydraethiops laevis Boulenger 1904
    Distribution: Cameroon

Hydraethiops melanogaster Gunther 1872
    Distribution: Gabon, Congo and Zaire
    Reference: Lanza and Vanni (1976)

Genus: LYCOPHIDION Fitzinger 1843
Species typica: *capense* Smith
Reference: Bourgeois (1968); Laurent (1968)

Lycophidion capense capense (Smith 1831)   *Lycodon*
    Distribution: Transvaal, Natal, Cape Province, Botswana, Namibia,
                Zambia and Zimbabwe
    Reference: Branch (1976); Broadley (1969)

Lycophidion capense jacksoni (Boulenger 1893)   *Lycophidium*
    Distribution: Southern Egypt, Sudan, western Ethiopia, Kenya,
                Tanzania, Uganda, Rwanda, Burundi and eastern Zaire

Lycophidion capense loveridgei Laurent 1968
    Distribution: Coastal Kenya and Tanzania

Lycophidion capense multimaculatum (Boettger 1888)   *Lycophidium*
    Distribution: Southern Zaire, Angola, Zambia and Zimbabwe
    Reference: Laurent (1956a) (1964b)

Lycophidion capense pembanum Laurent 1968
    Distribution: Pemba Island, Tanzania

Lycophidion capense vermiculatum Laurent 1968
    Distribution: Southeast Tanzania, northern Mozambique, Malawi and
                eastern Zimbabwe

Lycophidion depressirostre Laurent 1968
Distribution: Southern Sudan, Somalia, Kenya, Tanzania and Uganda

Lycophidion hellmichi Laurent 1964
Distribution: Southern Angola and Namibia

Lycophidion irroratum (Leach 1819)   *Coluber*
Distribution: Sierra Leone and Liberia east to Togo

Lycophidion laterale Hallowell 1857
Distribution: Senegal east to Congo

Lycophidion meleagre (Boulenger 1893)   *Lycophidium*
Distribution: Angola, southern Zaire and Tanzania

Lycophidion ornatum Parker 1936
Distribution: Angola, southern and eastern Zaire, Uganda, Rwanda,
        Burundi, western Tanzania and Sudan
Reference: Laurent (1956a)

Lycophidion polylepis Boulenger 1919
Distribution: Congo and Zaire

Lycophidion semiannule (Peters 1854)   *Lycophidium*
Distribution: Tanzania, Mozambique, Zambia, Zimbabwe and northern
        Zululand
Reference: Laurent (1964b)

Lycophidion semicinctum Dumeril and Bibron 1854
Distribution: Guinea Bissau east to Cameroon

Lycophidion uzungwense Loveridge 1932
Distribution: Tanzania

Lycophidion variegatum Broadley 1969
Distribution: Zimbabwe and adjacent South Africa
Reference: Branch (1976); Broadley (1974d)

Genus: MACROPROTODON Guichenot 1850
Species typica: *cucullatus* Geoffroy
Reference: Pasteur and Bons (1960); Steward (1971)

Macroprotodon cucullatus cucullatus (Geoffroy 1827)   *Coluber*
Distribution: Western Sahara and Saharan Morocco east to Egypt

Macroprotodon cucullatus brevis (Gunther 1862)   *Coronella*
    Distribution: Morocco

Genus: MEHELYA Csiki 1903
Species typica: *poensis* Smith
Reference: Bourgeois (1968); Loveridge (1939)

Mehelya capensis capensis (Smith 1847)   *Heterolepis*
    Distribution: Eastern South Africa, Zimbabwe, Malawi, Mozambique,
            Tanzania, Botswana, Namibia and Angola
    Reference: Dowling (1969); Fitzsimons (1974); Laurent (1956a)

Mehelya capensis fiechteri Scortecci 1929
    Distribution: Somalia
    Reference: Parker (1949)

Mehelya capensis savorgnani (Mocquard 1887)   *Heterolepis*
    Distribution: Southern Somalia, Kenya, southern Sudan, Uganda, Zaire,
            Congo, Cameroon, Gabon and Central African Republic
    Reference: Laurent (1956a); Loveridge (1957); Parker (1949)

Mehelya crossii (Boulenger 1895)   *Simocephalus*
    Distribution: Benin and Nigeria
    Reference: Dowling (1969)

Mehelya egbensis Dunger 1966
    Distribution: Nigeria

Mehelya guirali (Mocquard 1887)   *Heterolepis*
    Distribution: Liberia and Sierra Leone east to northern Zaire

Mehelya laurenti Witte 1959
    Distribution: Congo and adjacent Zaire

Mehelya nyassae (Gunther 1888)   *Simocephalus*
    Distribution: Kenya and Rwanda south through Mozambique, Malawi,
            Zambia and Zimbabwe to northeastern South Africa
    Reference: Dowling (1969); Langebartel (1968); Loveridge (1957)

Mehelya poensis (Smith 1847)   *Heterolepis*
    Distribution: Guinea and Sierra Leone east to Uganda and Angola
    Reference: Dowling (1969); Lanza and Vanni (1976); Laurent (1956a);
            Loveridge (1957); Saiff (1975)

Mehelya riggenbachi (Sternfeld 1910)   *Simocephalus*
    Distribution: Cameroon

Mehelya stenophthalmus (Mocquard 1887)   *Heterolepis*
    Distribution: Guinea Bissau east to Uganda and northern Zaire
    Reference: Laurent (1956a); Loveridge (1957)

Mehelya vernayi Bogert 1940
    Distribution: Angola and Namibia
    Reference: Fitzsimons (1974)

Subfamily Natricinae

Genus: AFRONATRIX Rossman and Eberle 1977
Species typica: *anoscopus* Cope

Afronatrix anoscopus (Cope 1861)   *Tropidonotus*
    Distribution: Liberia and Sierra Leone
    Reference: Cansdale (1961); Gartside and Dessauer (1977)
    Natrix firestonei Taylor and Weyer: Rossman (1976)

Genus: NATRIX Laurenti 1768
Species typica: *natrix* Linnaeus
Reference: Arnold and Burton (1978); Steward (1971)

Natrix maura (Linnaeus 1758)   *Coluber*
    Distribution: Morocco, Algeria and Tunisia
    Reference: Pasteur and Bons (1960)

Natrix natrix helvetica (Lacepede 1789)   *Coluber*
    Distribution: Morocco, Algeria and Tunisia
    Reference: Petter-Rousseaux (1953); Thorpe (1975a) (1975b) (1979)
        (1980)
    Natrix natrix astreptophora (Seoane); Pasteur and Bons (1960);
        Steward (1971)

Natrix tessellata tessellata (Laurenti 1768)   *Coronella*
    Distribution: Northeastern Egypt
    Reference: Gygax (1971); Marx (1968)

Subfamily Philothamninae

Genus: GASTROPYXIS Cope 1861
Species typica: *smaragdina* Schlegel

Gastropyxis smaragdina (Schlegel 1837)   *Dendrophis*
    Distribution: Uganda, Zaire and Angola west to Guinea Bissau
    Reference: Bourgeois (1968); Lanza and Vanni (1976)

Genus: HAPSIDOPHRYS Fischer 1856
Species typica: *lineatus* Fischer

Hapsidophrys lineatus Fischer 1856
    Distribution: Uganda, western Tanzania and Kenya west to Guinea
                Bissau
    Reference: Bourgeois (1968); Lanza and Vanni (1976)

Genus: PHILOTHAMNUS Smith 1840
Species typica: *semivariegata* Smith
Reference: Bourgeois (1968); Broadley (1959d); Fitzsimons (1974); Laurent
        (1956a); Loveridge (1958); Pitman (1974)

Philothamnus heterodermus heterodermus (Hallowell 1857)   *Chlorophis*
    Distribution: Angola, Zaire, Gabon, Congo, Cameroon, Nigeria, Togo,
                Ghana, Ivory Coast, Liberia, Sierra Leone, Guinea and
                Guinea Bissau

Philothamnus heterodermus carinatus (Andersson 1901)   *Chlorophis*
    Distribution: Uganda, Kenya, Zaire, Cameroon and Equatorial Guinea

Philothamnus heterodermus ruandae Loveridge 1951
    Distribution: Southwest Uganda, Rwanda, Burundi and adjacent Zaire

Philothamnus heterolepidotus (Gunther 1863)   *Ahaetulla*
    Distribution: Sudan, Uganda, Kenya, Tanzania, Mozambique, Angola,
                Zaire, Rwanda, Burundi, Cameroon, Nigeria, Togo, Benin
                and Gabon

Philothamnus hoplogaster (Gunther 1863)   *Ahaetulla*
   Distribution: Tanzania, Mozambique, Malawi, Zaire, Rwanda, Burundi,
              Zambia, Zimbabwe, Natal, Transvaal and Cape Province
   Reference: Sweeny (1971)

Philothamnus irregularis irregularis (Leach 1819)   *Coluber*
   Distibution: Southern Sudan and Ethiopia west to Senegal and south to
              Natal, then west to Namibia
   Reference: Broadley (1957a); Sweeny (1971)

Philothamnus irregularis battersbyi (Loveridge 1951)   *Chlorophis*
   Distribution: Ethiopia, Somalia, southern Sudan, northeast Zaire, eastern
              Uganda, Kenya and Tanzania

Philothamnus macrops (Boulenger 1895)   *Oligolepis*
   Distribution: Coastal Tanzania

Philothamnus natalensis (Smith 1840)   *Dendrophis*
   Distribution: Eastern Zimbabwe and southern Mozambique south to
              Cape Province
   Reference: Broadley (1966d); Jacobshagen (1920)

Philothamnus ornatus Bocage 1872
   Distribution: Southern Angola, Zambia and northern Zimbabwe

Philothamnus semivariegatus semivariegatus (Smith 1847)   *Dendrophis*
   Distribution: Ethiopia and southern Sudan south to Transvaal and
              Angola and west to Gambia, avoiding the forested areas
              of the Gulf of Guinea
   Reference: Sweeny (1971)

Philothamnus semivariegatus dorsalis (Bocage 1866)   *Leptophis*
   Distribution: Coastal Angola north along the coast to Gabon

Philothamnus semivariegatus girardi Bocage 1893
   Distribution: Annobon Island, Gulf of Guinea

Philothamnus semivariegatus nitidus (Gunther 1863)   *Ahaetulla*
   Distribution: Congo, eastern Gabon and Cameroon west to Ghana

Philothamnus semivariegatus thomensis Bocage 1882
   Distribution: Sao Toma Island, Gulf of Guinea

Subfamily Psammophinae

Genus: DROMOPHIS Peters 1869
Species typica: *praeornata* Schlegel
Reference: Bourgeois (1968); Loveridge (1940)

Dromophis lineatus (Dumeril and Bibron 1854)   *Dryophylax*
    Distribution: Guinea Bissau, Guinea, Liberia, Sierra Leone, Ivory Coast,
        Ghana, Togo, Benin, Nigeria, northern Cameroon,
        southern Chad, Central African Republic, Uganda,
        southern Sudan, Rwanda, Burundi, Tanzania, Malawi,
        eastern Zambia, Zimbabwe and Angola
    Reference: Fitzsimons (1974)

Dromophis praeornatus praeornatus (Schlegel 1837)   *Dendrophis*
    Distribution: Senegal east to Nigeria

Dromophis praeornatus gribinguiensis Angel 1921
    Distribution: Nigeria east to Central African Republic

Genus: HEMIRHAGERRHIS Boettger 1893
Species typica: *kelleri* Boettger
Reference: Bourgeois (1968); Parker (1949)

Hemirhagerrhis kelleri Boettger 1893
    Distribution: Ethiopia, Somalia and Kenya

Hemirhagerrhis nototaenia nototaenia (Gunther 1864)   *Coronella*
    Distribution: Southern Sudan, Somalia, Kenya, Tanzania, Mozambique,
        Malawi, Zambia, Zimbabwe, eastern Zaire, Botswana and
        Transvaal
    Reference: Fitzsimons (1974)

Hemirhagerrhis nototaenia viperinus (Bocage 1873)   *Psammophylax*
    Distribution: Southern Angola and Namibia
    Reference: Fitzsimons (1974)

Genus: PSAMMOPHIS Boie 1826
Species typica: *sibilans* Linnaeus
Reference: Bourgeois (1968); Broadley (1966c) (1975a) (1977b); Fitzsimons
        (1974); Loveridge (1940)

Psammophis aegyptius Marx 1958
    Distribution: Southern Egypt and Libya

Psammophis angolensis (Bocage 1872)   *Amphiophis*
    Distribution: Angola east through Shaba Province, Zaire, Zambia and
                Malawi to Mozambique, north to Tanzania and south
                through Zimbabwe to Transvaal; a relict population is
                present in Ethiopia
    Reference: Noble (1966)

Psammophis ansorgii Boulenger 1905
    Distribution: Central highlands of Angola
    Reference: Hellmich (1957)

Psammophis biseriatus biseriatus Peters 1881
    Distribution: Southern Somalia south through Kenya to northeastern
                Tanzania

Psammophis biseriatus tanganicus Loveridge 1940
    Distribution: Southern Libya, Sudan, Ethiopia, northern Somalia,
                northwest Kenya, Uganda and western Tanzania

Psammophis crucifer (Daudin 1803)   *Coluber*
    Distribution: Temperate regions of South Africa

Psammophis elegans (Shaw 1802)   *Coluber*
    Distribution: Senegal east to Nigeria

Psammophis jallae Peracca 1896
    Distribution: Southeast Angola, western Zambia, northeast Namibia,
                western and central Zimbabwe and western Transvaal

Psammophis leightoni leightoni Boulenger 1902
    Distribution: Southwestern Cape Province

Psammophis leightoni namibensis Broadley 1975
    Distribution: Southern Angola south through Namibia to western Cape
                Province

Psammophis leightoni trinasalis Werner 1902
    Distribution: Namibia, Botswana, northern Cape Province, Orange Free
                State and Transvaal

Psammophis notostictus Peters 1867
Distribution: Southwest Angola south to Cape Peninsula, east to
Albany District

Psammophis phillipsi (Hallowell 1844)   *Coluber*
Distribution: Senegal east to Kenya, south to Namibia, Botswana and
Swaziland

Psammophis pulcher Boulenger 1895
Distribution: Ethiopia

Psammophis punctulatus punctulatus Dumeril and Bibron 1854
Distribution: Sudan and Ethiopia

Psammophis punctulatus trivirgatus Peters 1878
Distribution: Uganda, Somalia, Kenya and northern Tanzania

Psammophis rukwae Broadley 1966
Distribution: Tanzania west across the Sudanese Savannah to Senegal

Psammophis schokari (Forskal 1775)   *Coluber*
Distribution: Western Sahara and Morocco east to Egypt and south to
Somalia
Reference: Domergue (1959b); Minton and Salanitro (1972); Pasteur
and Bons (1960)

Psammophis sibilans sibilans (Linnaeus 1758)   *Coluber*
Distribution: Morocco east to Egypt south to Somalia, Tanzania and
Malawi; west through Uganda, Central African Republic,
southern Chad and northern Nigeria to Senegal
Reference: Algauhari (1967); Broadley (1959a)

Psammophis sibilans brevirostris Peters 1881
Distribution: Transvaal, southeastern Botswana, northern Cape
Province, Swaziland, Natal, southwest Mozambique and
eastern and southwestern Zimbabwe

Psammophis sibilans leopardinus Bocage 1887
Distribution: Angola, northern Namibia and northern Zambia

Psammophis subtaeniatus subtaeniatus Peters 1881
Distribution: Southern Angola, northern Namibia, Botswana,
Zimbabwe, northern and eastern Transvaal, western
Mozambique and southern and southeastern Zambia
Reference: Broadley (1963c)

Psammophis subtaeniatus orientalis Broadley 1977
    Distribution: Eastern Kenya, Tanzania, Malawi and Mozambique

Psammophis trigrammus Gunther 1865
    Distribution: Southern Angola and Namibia

Genus: PSAMMOPHYLAX Fitzinger 1843
Species typica: *rhombeatus* Linnaeus
Reference: Bourgeois (1968); Broadley (1977a); Fitzsimons (1962) (1974)

Psammophylax rhombeatus rhombeatus (Linnaeus 1758)   *Coluber*
    Distribution: South Africa, western Swaziland and southern Namibia
    Reference: Chapman (1968); Radovanovic (1935)

Psammophylax rhombeatus ocellatus Bocage 1873
    Distribution: Southwestern Angola

Psammophylax tritaeniatus (Gunther 1868)   *Rhagerrhis*
    Distribution: Southern Tanzania and southeastern Zaire south through
                  Zambia, Malawi, Zimbabwe, Botswana and Mozambique
                  to South Africa, and west to northern Namibia and
                  southern Angola
    Reference: Broadley (1956); Chapman (1968); Laurent (1956a)

Psammophylax variabilis variabilis Gunther 1893
    Distribution: Southern Rwanda, Burundi, Zaire, southwestern Tanzania,
                  Malawi and adjacent Mozambique, Zambia and
                  northeastern Botswana

Psammophylax variabilis multisquamis (Loveridge 1932)
                                                        Trimerorhinus
    Distribution: Ethiopia, Kenya, northern Tanzania and northern Rwanda

Psammophylax variabilis vanoyei Laurent 1956
    Distribution: Monts Bleus, Zaire

Genus: RHAMPHIOPHIS Peters 1854
Species typica: *rostratus* Peters
Reference: Bourgeois (1968)

Rhamphiophis acutus acutus (Gunther 1888)   *Psammophis*
    Distribution: Angola, southern Zaire, western Tanzania and Burundi
    Reference: Broadley (1971a); Laurent (1956a)

Rhamphiophis acutus jappi Broadley 1971
Distribution: Western Zambia and northeastern Angola

Rhamphiophis acutus togoensis (Matschie 1893) *Psammophis*
Distribution: Northern Zaire and western Uganda west to Ghana and
Togo

Rhamphiophis acutus wittei Laurent 1956
Distribution: Southern Zaire, Angola and northern Zambia
Reference: Broadley (1971e); Laurent (1964b)

Rhamphiophis multimaculatus (Smith 1847) *Coronella*
Distribution: Namibia and Botswana
Reference: Fitzsimons (1974)

Rhamphiophis oxyrhynchus oxyrhynchus (Reinhardt 1843)
Psammophis
Distribution: Uganda, northeastern Zaire, Central African Republic,
Cameroon, Benin, Nigeria, Togo, Upper Volta, Mali,
Ivory Coast and Guinea
Reference: Laurent (1956a); Loveridge (1957)

Rhamphiophis oxyrhynchus rostratus Peters 1854
Distribution: Southern Sudan, Ethiopia, Somalia, Kenya, Tanzania,
Mozambique, Zaire, Malawi, Uganda, Zambia, Zimbabwe
and Transvaal
Reference: Fitzsimons (1974); Laurent (1956a); Loveridge (1957);
Parker (1949)

Rhamphiophis rubropunctatus (Fischer 1884) *Dipsina*
Distribution: Southern Sudan, Ethiopia, Somalia, eastern Kenya and
northern Tanzania
Reference: Loveridge (1957); Parker (1949)

Subfamily Pseudaspidinae

Genus: PSEUDASPIS Fitzinger 1843
Species typica: *canus* Linnaeus
Reference: Bourgeois (1968)

Pseudaspis cana cana (Linnaeus 1758)   *Coluber*
      Distribution: South Africa north to Angola and Kenya
      Reference: Brain (1959b) (1960); Branch (1973); Fitzsimons (1974);
                  Loveridge (1957); Schmidt and Inger (1957)

Pseudaspis cana anchietae (Bocage 1882)   *Ophirina*
      Distribution: Angola, Zaire, Rwanda and Burundi

Subfamily 'Uncertain'

Genus: ALLUAUDINA Mocquard 1894
Species typica: *bellyi* Mocquard
Reference: Guibe (1958)

Alluaudina bellyi Mocquard 1894
      Distribution: Madagascar

Alluaudina mocquardi Angel 1939
      Distribution: Madagascar

Genus: AMPLORHINUS Smith 1847
Species typica: *multimaculatus* Smith

Amplorhinus multimaculatus Smith 1847
      Distribution: South and eastern South Africa and eastern Zimbabwe
      Reference: Fitzsimons (1974)

Genus: COMPSOPHIS Mocquard 1894
Species typica: *albiventris* Mocquard
Reference: Guibe (1958)

Compsophis albiventris Mocquard 1894
      Distribution: Madagascar

Genus: DITYPOPHIS Gunther 1881
Species typica: *vivax* Gunther

Ditypophis vivax Gunther 1881
    Distribution: Socotra
    Reference: Parker (1949)

Genus: DROMICODRYAS Boulenger 1893
Species typica: *bernieri* Dumeril and Bibron
Reference: Guibe (1958)

Dromicodryas bernieri bernieri (Dumeril and Bibron 1854)   *Herpetodryas*
    Distribution: Madagascar

Dromicodryas bernieri ramavali Kaudern 1922
    Distribution: Madagascar

Dromicodryas quadrilineatus (Dumeril and Bibron 1854)   *Herpetodryas*
    Distribution: Madagascar

Genus: DUBERRIA Fitzinger 1826
Species typica: *lutrix* Linnaeus
Reference: Bourgeois (1968); Loveridge (1944c)

Duberria lutrix lutrix (Linnaeus 1758)   *Coluber*
    Distribution: South Africa and southern Mozambique
    Reference: Fitzsimons (1974)

Duberria lutrix abyssinica (Boulenger 1894)   *Homalosoma*
    Distribution: Ethiopia, Uganda, Kenya and northern Tanzania

Duberria lutrix atriventris (Sternfeld 1912)   *Homalosoma*
    Distribution: Rwanda, Burundi, southern Uganda, eastern Zaire and
                northwestern Tanzania
    Reference: Laurent (1956a)

Duberria lutrix currylindhali Laurent 1956
    Distribution: Kivu Province, Zaire

Duberria lutrix rhodesiana Broadley 1958
    Distribution: Eastern Zimbabwe
    Reference: Fitzsimons (1974)

Duberria lutrix shirana (Boulenger 1894)   *Homalosoma*
    Distribution: Malawi, western Mozambique, southern Tanzania and
            adjacent Zaire
    Reference: Laurent (1956a)

Duberria variegata (Peters 1854)   *Homalosoma*
    Distribution: Southern Mozambique south to Zululand
    Reference: Fitzsimons (1974); Haacke and Bruton (1978)

Genus: EIRENIS Jan 1863
Species typica: *collaris* Menetries

Eirenis africana (Boulenger 1914)   *Contia*
    Distribution: Sudan
    Reference: Parker (1949)

Genus: GRAYIA Gunther 1858
Species typica: *smythii* Leach
Reference: Laurent (1956a)

Grayia caesar (Gunther 1863)   *Xenurophis*
    Distribution: Equatorial Guinea, Cameroon, Congo and Zaire

Grayia ornata (Bocage 1866)   *Macrophis*
    Distribution: Angola north to Cameroon
    Reference: Lanza and Vanni (1976)

Grayia smythii (Leach 1818)   *Coluber*
    Distribution: Southern Sudan, Uganda, Kenya and Tanzania west to
            Angola and Senegal
    Reference: Lanza and Vanni (1976)

Grayia tholloni Mocquard 1897
    Distribution: Sudan, Uganda, Kenya, Tanzania, Zaire and Angola

Genus: HETEROLIODON Boettger 1913
Species typica: *torquatus* Boettger

Heteroliodon torquatus Boettger 1913
    Distribution: Madagascar
    Reference: Guibe (1958)

Genus: ITHYCYPHUS Gunther 1873
Species typica: *caudolineatus* Gunther = *goudoti* Schlegel
Reference: Guibe (1958)

Ithycyphus goudoti (Schlegel 1854)   *Herpetodryas*
    Distribution: Madagascar

Ithycyphus miniatus (Schlegel 1837)   *Coluber*
    Distribution: Madagascar and the Comoro Islands

Genus: LANGAHA Brugniere 1784
Species typica: *nasuta* Shaw
Reference: Guibe (1958)

Langaha alluaudi Mocquard 1901
    Distribution: Madagascar

Langaha nasuta Shaw 1790
    Distribution: Madagascar

Genus: LIMNOPHIS Gunther 1865
Species typica: *bicolor* Gunther
Reference: Laurent (1964b)

Limnophis bicolor bicolor Gunther 1865
    Distribution: Angola, Botswana and Zimbabwe
    Reference: Fitzsimons (1974)

Limnophis bicolor bangweolicus (Mertens 1936)   *Helicops*
    Distribution: Angola and Zaire

Genus: LIOHETERODON Dumeril and Bibron 1854
Species typica: *madagascariensis* Dumeril and Bibron
Reference: Guibe (1958)

Lioheterodon geayi Mocquard 1905
    Distribution: Madagascar

Lioheterodon madagascariensis (Dumeril and Bibron 1854)   *Heterodon*
    Distribution: Madagascar
    Reference: Campbell and Murphy (1977); Conant (1937); Saiff (1975)

Lioheterodon modestus (Gunther 1863)   *Heterodon*
    Distribution: Madagascar

Genus: LIOPHIDIUM Boulenger 1896
Species tyica: *trilineatum* Boulenger
Reference: Guibe (1958)

Liophidium mayottensis (Peters 1837)   *Ablades*
    Distribution: Comoro Islands

Liophidium rhodogaster (Schlegel 1837)   *Herpetodryas*
    Distribution: Madagascar

Liophidium torquatus (Boulenger 1888)   *Enicognathus*
    Distribution: Madagascar

Liophidium trilineatum Boulenger 1896
    Distribution: Madagascar

Liophidium vaillanti (Mocquard 1901)   *Idiophis*
    Distribution: Madagascar and the Mascarenes

Genus: LIOPHOLIDOPHIS Mocquard 1904
Species typica: *grandidieri* Mocquard
Reference: Guibe (1958)

Liopholidophis grandidieri Mocquard 1904
    Distribution: Madagascar

Liopholidophis lateralis (Dumeril and Bibron 1854)   *Leptophis*
    Distribution: Madagascar

Liopholidophis pinguis Parker 1925
    Distribution: Madagascar

Liopholidophis pseudolateralis Guibe 1954
    Distribution: Madagascar

Liopholidophis sexlineatus (Gunther 1882)   *Dromicus*
    Distribution: Madagascar

Genus: LYCODRYAS Gunther 1879
Species typica: *sanctijohannis* Gunther
Reference: Guibe (1958)

Lycodryas arctifasciatus (Dumeril and Bibron 1854)   *Heterurus*
    Distribution: Madagascar

Lycodryas betsileanus (Gunther 1880)   *Dipsas*
    Distribution: Madagascar

Lycodryas gaimardi gaimardi (Schlegel 1837)   *Dipsas*
    Distribution: Madagascar

Lycodryas gaimardi granuliceps (Boettger 1877)   *Dipsas*
    Distribution: Madagascar and the Comoro Islands

Lycodryas guentheri (Boulenger 1896)   *Stenophis*
    Distribution: Madagascar

Lycodryas inornatus (Boulenger 1896)   *Stenophis*
    Distribution: Madagascar

Lycodryas maculatus (Gunther 1858)   *Dipsadoboa*
    Distribution: Madagascar

Lycodryas sanctijohannis Gunther 1879
    Distribution: Comoro Islands

Lycodryas variabilis (Boulenger 1896)   *Stenophis*
    Distribution: Madagascar

Genus: LYCOGNATHOPHIS Boulenger 1893
Species typica: *seychellensis* Schlegel

Lycognathophis seychellensis (Schlegel 1837)   *Psammophis*
    Distribution: Seychelles

Genus: MADAGASCAROPHIS Mertens 1952
Species typica: *colubrina* Schlegel

Madagascarophis colubrina (Schlegel 1837)   *Dipsas*
    Distribution: Madagascar
    Reference: Campbell and Murphy (1977); Guibe (1958)

Genus: MICROPISTHODON Mocquard 1894
Species typica: *ochraceus* Mocquard

Micropisthodon ochraceus Mocquard 1894
    Distribution: Madagascar
    Reference: Guibe (1958)

Genus: MIMOPHIS Gunther 1868
Species typica: *mahafalensis* Grandidier

Mimophis mahafalensis (Grandidier 1867)   *Psammophis*
    Distribution: Madagascar
    Reference: Guibe (1958)

Genus: NATRICITERES Loveridge 1953
Species typica: *olivacea* Peters
Reference: Bourgeois (1968); Broadley (1966b); Loveridge (1958)

Natriciteres fuliginoides (Gunther 1858)   *Coronella*
    Distribution: Guinea east to Zaire

Natriciteres olivacea (Peters 1854)   *Coronella*
    Distribution: Sudan south to Mozambique and west to Angola and
                Guinea
    Reference: Lanza and Vanni (1976)

Natriciteres variegata variegata (Peters 1861)   *Mizodon*
    Distribution: Sierra Leone east to Congo and northern Zaire
    Reference: Lanza and Vanni (1976)

Natriciteres variegata bipostocularis Broadley 1962
    Distribution: Southern Zaire, Angola and northern Zambia

Natriciteres variegata pembana (Loveridge 1935)   *Natrix*
    Distribution: Pemba Island, Tanzania

Natriciteres variegata sylvatica Broadley 1966
    Distribution: Southern Tanzania, western Mozambique, Malawi and
                eastern Zimbabwe

Genus: PARARHADINEA Boettger 1898
Species typica: *melanogaster* Boettger

Pararhadinea melanogaster Boettger 1898
    Distribution: Madagascar
    Reference: Guibe (1958)

Genus: POECILOPHOLIS Boulenger 1903
Species typica: *cameronensis* Boulenger

Poecilopholis cameronensis Boulenger 1903
    Distribution: Cameroon

Genus: PROSYMNA Gray 1849
Species typica: *meleagris* Reinhardt
Reference: Bourgeois (1968); Broadley (1979); Loveridge (1958)

Prosymna ambigua ambigua Bocage 1873
    Distribution: Northern Angola, Zaire and Congo
    Reference: Laurent (1956a)

Prosymna ambigua bocagii Boulenger 1897
    Distribution: Sudan, Uganda, northern Zaire and Central African
                Republic
    Reference: Laurent (1956a); Pitman (1974)

Prosymna ambigua brevis Laurent 1954
    Distribution: Northeastern Angola and adjacent Zaire
    Reference: Laurent (1964b)

Prosymna ambigua ornatissima Barbour and Loveridge 1928
    Distribution: Uluguru Mountains, Tanzania

Prosymna ambigua ruspolii (Boulenger 1896)   *Asthenophis*
    Distribution: Somalia

Prosymna ambigua stuhlmanni (Pfeffer 1893)   *Ligonirostra*
    Distribution: Kenya, Tanzania, Mozambique, Malawi, Zambia,
                Zimbabwe, Rwanda, Burundi, southeastern Zaire and
                Transvaal
    Reference: Fitzsimons (1974); Sweeny (1971)

Prosymna angolensis Boulenger 1915
    Distribution: Namibia and southwestern Angola
    Reference: Fitzsimons (1974)

Prosymna bivittata Werner 1903
  Distribution: Namibia, Botswana, western Zimbabwe, Transvaal and
        southern Mozambique
  Reference: Broadley (1965c); Fitzsimons (1974)

Prosymna frontalis (Peters 1867)   *Temnorhynchus*
  Distribution: Namibia
  Reference: Fitzsimons (1974)

Prosymna greigerti Mocquard 1906
  Distribution: Lobi Region, Upper Volta

Prosymna janii Bianconi 1862
  Distribution: Mozambique and Natal
  Reference: Fitzsimons (1974); Haacke and Bruton (1978)

Prosymna meleagris meleagris (Reinhardt 1843)   *Calamaria*
  Distribution: Guinea east to Nigeria

Prosymna meleagris laurenti Loveridge 1958
  Distribution: Uganda, northern Zaire and southern Sudan west to
        Mauritania
  Reference: Pitman (1974)

Prosymna pitmani Battersby 1951
  Distribution: Southeastern Tanzania

Prosymna somalica Parker 1930
  Distribution: Somalia

Prosymna sundevalli sundevalli (Smith 1849)   *Temnorhynchus*
  Distribution: Transvaal, Orange Free State, Lesotho, Natal and Cape
        Province
  Reference: Broadley (1965c); Fitzsimons (1974)

Prosymna sundevalli lineata (Peters 1871)   *Temnorhynchus*
  Distribution: Mozambique, Zimbabwe, eastern Botswana, Zululand,
        eastern and northern Transvaal
  Reference: Broadley (1965c); Fitzsimons (1974)

Genus: PSEUDOTARBOPHIS Domergue 1955
Species typica: *gabesiensis* Domergue

Pseudotarbophis gabesiensis Domergue 1955
 Distribution: Tunisia
 Reference: Domergue (1956b)

Genus: PSEUDOXYRHOPUS Gunther 1881
Species typica: *microps* Gunther
Reference: Guibe (1958)

Pseudoxyrhopus ambreensis Mocquard 1894
 Distribution: Madagascar

Pseudoxyrhopus dubius Mocquard 1904
 Distribution: Madagascar

Pseudoxyrhopus heterurus (Jan 1893) *Homalocephalus*
 Distribution: Madagascar

Pseudoxyrhopus imerinae (Gunther 1890) *Liophis*
 Distribution: Madagascar

Pseudoxyrhopus microps Gunther 1881
 Distribution: Madagascar

Pseudoxyrhopus occipitalis Boulenger 1896
 Distribution: Madagascar

Pseudoxyrhopus punctatus (Peters 1880) *Liophis*
 Distribution: Madagascar

Pseudoxyrhopus quinquelineatus (Gunther 1881) *Liophis*
 Distribution: Madagascar

Pseudoxryhopus tritaeniatus Mocquard 1894
 Distribution: Madagascar

Genus: PYTHONODIPSAS Gunther 1868
Species typica: *carinata* Gunther

Pythonodipsas carinata Gunther 1868
 Distribution: Namibia and southern Angola
 Reference: Fitzsimons (1974); Stuart (1976)

Genus: SCAPHIOPHIS Peters 1870
Species typica: *albopunctatus* Peters
Reference: Bourgeois (1968)

Scaphiophis albopunctatus albopunctatus Peters 1870
  Distribution: Uganda, Kenya, Tanzania and southern Sudan west to
      Guinea and Sierra Leone

Scaphiophis albopunctatus raffreyi Bocourt 1875
  Distribution: Ethiopia
  Reference: Parker (1949)

Chapter 18

FAMILY ELAPIDAE

Order: Serpentes
Suborder: Alethinophidia
Superfamily: Elapsoidea
Family: Elapidae
Subfamily: Bungarinae
Tribe: Dendroaspini

Genus: DENDROASPIS Schlegel 1848
Species typica: *jamesonii* Traill
Reference: Bourgeois (1968)

Dendroaspis angusticeps (Smith 1849)   *Naja*
    Distribution: Kenya south to Natal including Malawi and eastern
            Zimbabwe
    Reference: Phisalix (1914) (1922)

Dendroaspis jamesoni jamesoni (Traill 1843)   *Elaps*
    Distribution: Guinea, Liberia and Upper Volta east to Zaire and Central
            African Republic
    Reference: Kyegombe and others (1978); Laurent (1964b); Patel and
            Excell (1974) (1975); Raina and others (1977a)

Dendroaspis jamesoni kaimosae Loveridge 1936
    Distribution: Western Kenya, Uganda, Rwanda and adjacent Zaire
    Reference: Joubert, Strydom and Taljaard (1978); Leloup (1964)

185

Dendroaspis polylepis polylepis Gunther 1864
Distribution: Kenya south to Swaziland, Transvaal and Lesotho, west
to include Uganda, Zaire, Angola and Botswana
Reference: Broadley (1961c); Joubert and Strydom (1978); Schmidt and
Inger (1957); Strydom (1972) (1977a) (1977b)

Dendroaspis polylepis antinori Peters 1873
Distribution: Somalia, Ethiopia, northeastern Uganda and northern
Kenya
Reference: Parker (1949)

Dendroaspis viridis (Hallowell 1844)   *Leptophis*
Distribution: Gambia, Senegal, Guinea, Guinea Bissau, Liberia, Ivory
Coast, Ghana and Sierra Leone
Reference: Banks, Miledi and Shipolini (1974); Eigenberger (1928)

Subfamily: Bungarinae
Tribe: Najini

Genus: ASPIDELAPS Fitzinger 1843
Species typica: *lubrica* Laurenti

Aspidelaps lubricus lubricus (Laurenti 1768)   *Natrix*
Distribution: Cape Province and Orange Free State

Aspidelaps lubricus cowlesi Bogert 1940
Distribution: Southern Angola

Aspidelaps lubricus infuscatus Mertens 1954
Distribution: Namibia

Aspidelaps scutatus scutatus (Smith 1849)   *Cryptophis*
Distribution: Namibia, Botswana, southwest Zimbabwe and northwest
Transvaal
Reference: Broadley (1968c)

Aspidelaps scutatus fulafulus (Bianconi 1849)   *Naia*
Distribution: Southeastern Zimbabwe and southern Mozambique
Reference: Broadley (1968c)

Aspidelaps scutatus intermedius Broadley 1968
Distribution: Eastern Transvaal

Genus: BOULENGERINA Dollo 1886
Species typica: *stormsi* Dollo

Boulengerina annulata annulata (Buchholz and Peters 1877)   *Naja*
 Distribution: Cameroon, Gabon, Congo and Zaire

Boulengerina annulata stormsi Dollo 1886
 Distribution: The shores of Lake Tanganyika
 Reference: Kratzer (1965); Loveridge (1931a); Witte (1962)

Boulengerina christyi Boulenger 1904
 Distribution: Zaire and Congo

Genus: ELAPSOIDEA Bocage 1866
Species typica: *guentheri* Bocage
Reference: Broadley (1971c)

Elapsoidea chelazzii Lanza 1979
 Distribution: Somalia

Elapsoidea guentheri Bocage 1866
 Distribution: Southern Congo, Zaire, Angola, Zambia and Zimbabwe

Elapsoidea laticincta (Werner 1919)   *Elapechis*
 Distribution: Southern Sudan, northern Zaire and Central African
        Republic

Elapsoidea loveridgei loveridgei Parker 1949
 Distribution: Kenya and northern Tanzania

Elapsoidea loveridgei colleti Laurent 1956
 Distribution: Eastern Zaire, southwestern Uganda and Rwanda

Elapsoidea loveridgei multicincta Laurent 1956
 Distribution: Northeastern Zaire, Uganda, western Kenya, northern
        Tanzania and southern Ethiopia

Elapsoidea loveridgei scalaris Laurent 1960
 Distribution: Kivu Province, Zaire

Elapsoidea nigra Gunther 1888
 Distribution: Northeastern Tanzania

Elapsoidea semiannulata semiannulata Bocage 1882
Distribution: Northern Namibia, Angola, southwestern Zaire and
western Zambia

Elapsoidea semiannulata boulengeri Boettger 1895
Distribution: Tanzania south to Mozambique west to Zambia, Botswana
and northern Cape Province

Elapsoidea semiannulata moebiusi Werner 1897
Distribution: Zaire, Central African Republic and Chad west to Senegal
and Mauritania

Elapsoidea sundevallii sundevallii (Smith 1848)   *Elaps*
Distribution: Natal, Swaziland and southeastern Transvaal
Reference: Fitzsimons (1962); McDowell (1968)

Elapsoidea sundevallii decosteri Boulenger 1888
Distribution: Southern Mozambique and northern Zululand

Elapsoidea sundevallii fitzsimonsi Loveridge 1944
Distribution: Namibia, western Botswana and northern Cape Province
Reference: Fitzsimons and Brain (1958)

Elapsoidea sundevallii longicauda Broadley 1971
Distribution: Northern Transvaal, southeastern Zimbabwe and adjoining
Mozambique

Elapsoidea sundevallii media Broadley 1971
Distribution: Transvaal and Orange Free State west to northern Cape
Province

Genus: HEMACHATUS Fleming 1822
Species typica: *haemachata* Lacepede

Hemachatus haemachatus (Lacepede 1789)   *Coluber*
Distribution: South Africa, Swaziland and Zimbabwe
Reference: Barzilay, Kaminsky and Condrea (1978); Bengis and Noble
(1976); Duvernoy (1832); Fryklund and Eaker (1973);
Hokama and others (1976); Joubert and Taljaard (1980);
Pringle (1954); Stejneger (1936); Strydom and Botes (1971);
Willemse and Hattingh (1979)

Genus: NAJA Laurenti 1768
Species typica: *naja* Linnaeus
Reference: Broadley (1968a) (1974c); Harding and Welch (1980)

Naja haje haje (Linnaeus 1758)   *Coluber*
Distribution: Saharan Morocco east to Egypt then south to Kenya and
        Tanzania; west from Sudan and northeastern Zaire to
        Senegal
Reference: Behler and Brazaitis (1974); Broadley (1958b); Hager
        (1905); Kopeyan and others (1973); Joubert and Taljaard
        (1978a) (1978b) (1978c); Miranda and others (1970);
        Mohamed and others (1975); Parker (1949); Warrell, Barnes
        and Piburn (1976)

Naja haje anchietae Bocage 1879
Distribution: Southern Angola, northern Namibia, Botswana,
        northwestern Zimbabwe and western Zambia

Naja haje annulifera Peters 1854
Distribution: Zimbabwe and eastern Transvaal south to Zululand;
        Zambia, Malawi and northern Mozambique
Reference: Joubert (1977a) (1977b)

Naja melanoleuca Hallowell 1857
Distribution: Senegal east to western Ethiopia and southern Somalia,
        south to Angola and Natal
Reference: Botes (1972); Carlsson (1974); Carlsson and Louw (1978);
        Joubert and Van Der Walt (1975); Langebartel (1968); Lee
        (1976); Poilleux and Boquet (1972); Shipolini, Bailey and
        Banks (1974); Wilson (1959)

Naja mossambica mossambica Peters 1854
Distribution: Southeastern Tanzania south to Natal, west through
        Namibia to Angola
Reference: Gregoire and Rochat (1977); Joubert (1977c); Louw (1974a)
        (1974b) (1974c); Martin-Moutet and Rochat (1979)

Naja mossambica katiensis Angel 1922
Distribution: Mali, northern Nigeria and northern Ghana

Naja mossambica pallida Boulenger 1896
Distribution: Southern Egypt south to northern Tanzania and Somalia
Reference: Greenham (1978); Raina, Nganga and Telang (1978); Raina
        and others (1977b)

Naja nigricollis nigricollis Reinhardt 1843
Distribution: Senegal east to Kenya, Tanzania, Zambia and Nambia
Reference: Boquet and others (1966a) (1966b); Chippaux and others
(1978); Detrait and Boquet (1972); Detrait, Izard and Boquet
(1959) (1960); Dumarey and Boquet (1972); Dupont (1977);
Kopeyan and others (1973); Lee, Ho and Eaker (1977);
Mohamed and Nawar (1975); Wahlstrom (1971); Warrell
and Ormerod (1976); Warrell and others (1976)

Naja nigricollis nigricincta Bogert 1940
Distribution: Northern Namibia and adjoining Angola
Reference: Boycott and Haacke (1979)

Naja nigricollis woodi Pringle 1955
Distribution: Southern Namibia and western Cape Province
Reference: Boycott and Haacke (1979)

Naja nivea (Linnaeus 1758)   *Coluber*
Distribution: Namibia, Botswana and South Africa
Reference: Botes and Viljoen (1976); Botes and others (1971); Earl and
Excell (1972); Hokama and others (1976); Madsen,
Ludstrom and Fohlman (1979); Meij and Meyer (1977)

Genus: PARANAJA Loveridge 1944
Species tyica: *anomala* Sternfeld

Paranaja multifasciata multifasciata (Werner 1902)   *Naia*
Distribution: Northern Zaire and Congo

Paranaja multifasciata anomala (Sternfeld 1917)   *Naia*
Distribution: Cameroon

Genus: PSEUDOHAJE Gunther 1858
Species typica: *nigra* Gunther
Reference: Hughes (1976)

Pseudohaje goldii (Boulenger 1895)   *Naia*
Distribution: Ghana east to Kenya and Angola

Pseudohaje nigra Gunther 1858
    Distribution: Sierra Leone and Liberia east to Nigeria

Genus: WALTERINNESIA Lataste 1887
Species typica: *aegyptia* Lataste

Walterinnesia aegyptia Lataste 1887
    Distribution: Egypt
    Reference: Kochva (1962); Lee, Chen and Mebs (1976); Schmidt and
            Inger (1957)

Chapter 19

FAMILY HYDROPHIIDAE

Order: Serpentes
Suborder: Alethinophidia
Infraorder: Caenophidia
Superfamily: Elapsoidea
Family: Hydrophiidae
Subfamily: Hydrophiinae

Genus: ENHYDRINA Gray 1849
Species typica: *schistosus* Daudin

Enhydrina schistosa (Daudin 1803)   *Hydrophis*
    Distribution: Madagascar
    Reference: Guibc 1958

Genus: PELAMIS Daudin 1803
Species typica: *platura* Linnaeus

Pelamis platurus (Linnaeus 1766)   *Anguis*
    Distribution: Eastern coastal waters of Africa
    Reference: Bolanos and others (1974); Burns and Pickwell (1972);
                Dunson (1968); Dunson and Ehlert (1971); Fitzsimons
                (1974); Graham (1974a) (1974b); Graham, Gee and Robison
                (1975); Graham, Rubinoff and Hecht (1971); Greene (1973);
                Guibe (1958); Hecht, Kropach and Hecht (1974); Hibbard
                and LaVergne (1972); Klawe (1964); Kropach (1971) (1972)
                (1975); Pickwell (1971); Shipman and Pickwell (1973); Tu,
                Lin and Bieber (1975); Visser (1967); Zeiller (1969)

Chapter 20

FAMILY VIPERIDAE

Order: Serpentes
Suborder: Alethinophidia
Infraorder: Caenophidia
Superfamily: Viperoidea
Family: Viperidae
Subfamily: Viperinae

Genus: ADENORHINOS Marx and Rabb 1965
Species typica: *barbouri* Loveridge

Adenorhinos barbouri (Loveridge 1930)   *Atheris*
    Distribution: Uzungwe and Ukinga Mountains, Tanzania

Genus: ATHERIS Cope 1862
Species typica: *chloroechis* Schlegel

Atheris ceratophorus Werner 1895
    Distribution: Usambara Mountains, Tanzania

Atheris chloroechis (Schlegel 1855)   *Vipera*
    Distribution: Guinea and Sierra Leone east to Cameroon

Atheris desaixi Ashe 1968
    Distribution: near Chuka, Kenya

Atheris hindii (Boulenger 1910)   *Vipera*
   Distribution: Kinangop and Aberdare Mountains, Kenya
   Reference: Andren (1976); Ionides and Pitman (1965)

Atheris hispidus Laurent 1955
   Distribution: Zaire, Uganda and western Kenya

Atheris katangensis Witte 1953
   Distribution: Zaire

Atheris nitschei nitschei Tornier 1902
   Distribution: Western Tanzania, Rwanda, Burundi, Uganda and adjacent
            Zaire
   Reference: Kochva (1962)

Atheris nitschei rungweensis Bogert 1940
   Distribution: Southwest Tanzania, northern Malawi and northeastern
            Zambia

Atheris squamiger (Hallowell 1854)   *Echis*
   Distribution: Western Kenya and Uganda west to Cameroon and south
            to Angola
   Reference: Kochva (1962); Vogel (1964)
   Atheris squamiger anisolepis Mocquard: Harding and Welch (1980)
   Atheris squamiger robustus Laurent: Harding and Welch (1980)

Atheris superciliaris (Peters 1854)
   Distribution: Mozambique, Malawi and southern Tanzania

Genus: BITIS Gray 1842
Species typica: *arietans* Merrem

Bitis arietans arietans (Merrem 1820)   *Vipera*
   Distribution: South Africa north to Kenya in the east and Morocco in
            the west
   Reference: Bolt and Ewer (1964); Broadley and Parker (1976); Howard
            (1975); Janecek (1976); Kochva (1962); Mertens (1968); Otis
            (1973); Parsons and Cameron (1977); Peters and Broadley
            (1967); Radovanovic (1935); Schmidt and Inger (1957);
            Van der Walt (1972); Van der Walt and Joubert (1971)
            (1972); Vogel (1964); Warrell, Ormerod and Davidson
            (1975); Willemse, Hattingh and Coetzee (1979); Willemse
            and others (1979)

Bitis arietans somalica Parker 1949
    Distribution: Somalia and northern Kenya

Bitis atropos atropos (Linnaeus 1758)   *Coluber*
    Distribution: Coastal southern Cape Province northeast through
                Lesotho, Orange Free State to eastern Zimbabwe
    Reference: Hurwitz and Hull (1971); Montgomery (1959); Reitz (1977)

Bitis atropos unicolor Fitzsimons 1959
    Distribution: Southeastern Transvaal north of Belfast

Bitis caudalis (Smith 1849)   *Vipera*
    Distribution: Namibia and southern Angola east to Transvaal and
                Zimbabwe
    Reference: Broadley (1972b); Haacke (1975); Kochva (1962)

Bitis cornuta cornuta (Daudin 1803)   *Vipera*
    Distribution: Western Cape Province north to Damaraland, Namibia
    Reference: Haacke (1975); Jacobshagen (1920); Kochva (1962)

Bitis cornuta inornata (Smith 1849)   *Echidna*
    Distribution: Matjesfontein east to Albany District, southern Karoo,
                South Africa
    Reference: Underwood (1968); Visser and Chapman (1978)
    Bitis cornuta albanica Hewitt: Harding and Welch (1980); Visser and
                Chapman (1978)

Bitis gabonica (Dumeril and Bibron 1845)   *Echidna*
    Distribution: Zululand and Angola north to Sudan and Guinea
    Reference: Akester (1979); Botes and Viljoen (1974a) (1974b); Broadley
                and Parker (1976); Forbes and others (1969); Gaffney,
                Marsh and Whaler (1973); Grasset and Zoutendyk (1938);
                Huffman (1974); Marsh and Whaler (1974); Mebs (1969)
                (1970); Phisalix (1914); Schmidt and Inger (1957); Snedigar
                and Rokosky (1949); Staley (1929); Vandeventer and Schmidt
                (1977); Viljoen and Botes (1979); Viljoen, Botes and Schabort
                (1975); Viljoen, Meehan and Botes (1979); Viljoen, Schabort
                and Botes (1974); Viljoen, Visser and Botes (1976) (1977);
                Whaler (1971) (1972) (1975)
    Bitis gabonica rhinoceros (Schlegel); Harding and Welch (1980)

Bitis heraldica (Bocage 1889)   *Vipera*
    Distribution: Central Angola

Bitis nasicornis (Shaw 1802)   *Coluber*
    Distribution: Southern Sudan, western Kenya and Uganda west through
                Zaire and Angola to Guinea
    Reference: Kochva (1962); Mackay, Ferguson and McNicol (1970);
                Marsh and Glatston (1974); Schmidt and Inger (1957)

Bitis peringueyi (Boulenger 1888)   *Vipera*
    Distribution: Namibia and southern Angola
    Reference: Haacke (1975)

Bitis schneideri (Boettger 1886)   *Vipera*
    Distribution: Namibia
    Reference: Haacke (1975)
    Bitis caudalis paucisquamata Mertens: Harding and Welch (1980)

Bitis worthingtoni Parker 1932
    Distribution: Kenya

Bitis xeropaga Haacke 1975
    Distribution: Namibia south to Cape Province

Genus: CERASTES Laurenti 1768
Species typica: *cerastes* Linnaeus
Reference: Harding and Welch (1980); Hassan and El Hawary (1977);
        Kochva (1962); Labib, Halim and Farag (1979); Warburg (1964)

Cerastes cerastes cerastes (Linnaeus 1758)   *Coluber*
    Distribution: Egypt west to Saharan Morocco, Mauritania, Mali and
                Niger
    Reference: Mohamed, El-Serougi and Khaled (1969); Mohamed and
              Khaled (1966); Radovanovic (1935); Saint Girons and Saint
              Girons (1956)

Cerastes vipera (Linnaeus 1758)   *Coluber*
    Distribution: Egypt west to Saharan Morocco, Mauritania, Mali and
                Niger
    Reference: Heatwole and Davidson (1976); Langebartel (1968); Vogel
              (1964)

Genus: ECHIS Merrem 1820
Species typica: *carinata* Schneider
Reference: Harding and Welch (1980)

Echis carinatus (Schneider)
Reference: Bhattacharya and Gaitonde (1979); Cheymol and others
(1973); Detrait, Izard and Boquet (1960); Izard and Boquet
(1958); Kornalik (1963); Kornalik and Taborska (1972);
Lefrou and Martignoles (1954); Morita, Iwanga and Suzuki
(1976); Parsons and Cameron (1977); Poguda (1972); Reid
(1977); Schiek, Kornalik and Habermann (1972); Somani
and Arora (1962); Taborska (1971);Theakston, Lloyd-Jones
and Reid (1977); Vogel (1964); Warrell and Arnett (1976);
Warrell and others (1974) (1977); Zimmermann, Habermann
and Lasch (1971)

Echis carinatus leakeyi Stemmler and Sochurek 1969
Distribution: Kenya, Somalia and southern Ethiopia

Echis carinatus ocellatus Stemmler 1970
Distribution: West Africa

Echis carinatus pyramidum (Geoffroy St. Hilaire 1827)   *Scythale*
Distribution: Mauritania and Saharan Morocco east to Egypt, Sudan and
northern Ethiopia

Echis coloratus Gunther 1878
Distribution: Egypt east of the Nile; Socotra
Reference: Anderson (1900); Warburg (1964)

Genus: VIPERA Laurenti 1768
Species typica: *aspis* Linnaeus
Reference: Harding and Welch (1980)

Vipera lastastei gaditana Saint Girons 1977
Distribution: Mediterranean coast of Morocco, Algeria and Tunisia
Reference: Saint Girons (1978)

Vipera lastastei monticola Saint Girons 1953
Distribution: Haut Atlas, Morocco
Reference: Saint Girons (1978); Saint Girons and Detrait (1978);
Steward (1971)

Vipera lebetina (Linnaeus)
Reference: Arnold and Burton (1978); Eggertsen, Fohlman and Sjoquist
(1980); Kochva (1962); Poguda (1972); Reymond (1956);
Steward (1971)

Vipera lebetina deserti Anderson 1892
    Distribution· Saharan Algeria, Tunisia and Libya
    Reference: Saint Girons and Detrait (1978)

Vipera lebetina mauritanica (Gray 1849)   *Clotho*
    Distribution: Northern Morocco, Algeria, Tunisia and Libya

Subfamily Causinae

Genus: CAUSUS Wagler 1830
Species typica: *rhombeata* Lichtenstein

Causus bilineatus Boulenger 1905
    Distribution: Angola, Zambia, Zaire and Rwanda

Causus defilippii (Jan 1862)   *Heterodon*
    Distribution: Tanzania south through Malawi, Zambia, Zimbabwe,
                Mozambique to Transvaal and Natal
    Reference: Fitzsimons (1974); Haas (1938) (1952)

Causus lichtensteini (Jan 1859)   *Aspidelaps*
    Distribution: Kenya and Uganda west to Liberia
    Reference: Haas (1938) (1953)

Causus maculatus (Hallowell 1842)   *Distichurus*
    Distribution: Senegal east to Chad, Zaire and Angola
    Reference: Hughes (1978); Warrell, Ormerod and Davidson (1976)

Causus resimus (Peters 1862)   *Heterophis*
    Distribution: Sudan, Ethiopia, Somalia, Kenya, Uganda, Tanzania,
                Mozambique, Malawi, Zambia, Rwanda, Burundi, eastern
                Zaire and Chad
    Reference: Haas (1938) (1952); Langebartel (1968)

Causus rhombeatus (Lichtenstein 1823)   *Sepedon*
    Distribution: Angola and Natal north to Sudan and Somalia in the east
                and Senegal and Mauritania in the west
    Reference: Fitzsimons (1974); Haas (1931a) (1931b) (1938) (1952);
                Kochva (1962); Phisilax (1914) (1922); Pringle (1954);
                Radovanovic (1935) (1967); Shayer-Wollberg and Kochva
                (1967); Sulter (1962); Woodward (1933)

Appendix 1

FAMILIAL REFERENCES

Amphisbaenidae: Gans (1967); Loveridge (1941c).

Trogonophidae: Gans (1967); Loveridge (1941c).

Gekkonidae: Loveridge (1942a) (1947); Underwood (1954); Wermuth (1965).

Agamidae: Wermuth (1967.

Chamaeleonidae: Mertens (1966).

Scincidae: Greer (1970a) (1970b); Mittleman (1954).

Lacertidae: Arnold (1973); Bons (1959); Boulenger (1920) (1921).

Cordylidae: Loveridge (1942b) (1944b); Wermuth (1968).

Anguidae: Wermuth (1969).

Varanidae: Mertens (1963).

Leptotyphlopidae: Hahn (1978a); List (1966); McDowell (1967).

Typhlopidae: Hahn (1978a); List (1966); McDowell (1967) (1974).

Boidae: McDowell (1975) (1979); Stimson (1969).

Colubridae: Bourgeois (1968); Underwood (1967).

Atractaspidinae: Harding and Welch (1980); Laurent (1950b).

Elapidae: Harding and Welch (1980).

Hydrophiidae: Burger and Natsuno (1974); Harding and Welch (1980); Smith (1926); Voris (1977).

Viperidae: Harding and Welch (1980); Marx and Rabb (1965).

Appendix 2

REGIONAL REFERENCES

ALDABRA: see Madagascar; ALGERIA: Angel and Lhote (1938), Domergue (1959a), Doumergue (1901); ANGOLA: Fitzsimons (1943) (1974), Laurent (1964b); BENIN: Angel (1933c), Villiers (1958); BOTSWANA: Fitzsimons (1943) (1974); BURUNDI: Laurent (1956), Witte (1962); CAMEROON: Muller (1910), Sternfeld (1908a); CHAD: Angel and Lhote (1938), Tercafs (1962); COMORO ISLANDS: see Madagascar; EGYPT: Anderson (1898), Marx (1968); EUROPA ISLAND: see Madagascar; GABON: Knoepffler (1968); GHANA: Cansdale (1951), Hoogmoed (1973a) (1973b) (1974), Hughes and Barry (1969), Lesson (1950), Leston (1970), Leston and Hughes (1968), Orton and Morrison (1946); GUINEA BISSAU: Manacas (1951) (1955); IVORY COAST: Barbault (1967) (1971b) (1974c) (1974e) (1975), Doucet (1963), Roux-Esteve (1969); JUAN DE NOVA ISLAND: see Madagascar; KENYA: Loveridge (1957); LIBERIA: Briscoe (1949), Loveridge (1941d); LIBYA: Angel and Lhote (1938), Kramer and Schnurrenberger (1963); MADAGASCAR: Angel (1930b) (1942), Blanc (1971), Blanc and Blanc (1967), Guibe (1958); MALAWI: Stewart (1968), Sweeney (1960) (1971); MALI: Angel (1933); MASCARENE ISLANDS: Vinson and Vinson (1969), see also Madagascar; MAURITANIA: Angel (1933c) (1938a), Angel and Lhote (1938), Villiers (1950a); MOROCCO: Aellen (1951), Bons (1957) (1958) (1959), Domergue (1959a) (1959b), Pasteur and Bons (1960); MOZAMBIQUE: Fitzsimons (1943) (1974), Loveridge (1953), Manacas (1952) (1957); NAMIBIA: Fitzsimons (1938) (1943) (1974), Mertens (1955); NIGER: Angel (1933c), Angel and Lhote (1938), Villiers (1950b); NIGERIA: Dunger (1967a) (1967b) (1967c) (1968a) (1968b) (1971a) (1971b) (1972a) (1972b) (1973); REPUBLIC OF SOUTH AFRICA: Fitzsimons (1943) (1962) (1966) (1974), Pienaar (1966), Visser and Chapman (1978); RWANDA: Laurent (1956), Witte (1962); SENEGAL: Angel (1933c); SEYCHELLES: Rendahl (1939), Vesey-Fitzgerald (1948), see also Madagascar; SIERRA LEONE: Aylmer (1922), Menzies

(1966); SOMALIA: Gans, Laurent and Pandit (1965), Parker (1942) (1949); SWAZILAND: Fitzsimons (1943) (1974); TANZANIA: Loveridge (1957); TOGO: Hulselmans, Roo and Vree (1970) (1971), Sternfeld (1908b); TUNISIA: Blanc (1936), Domergue (1959a) (1959b); UGANDA: Loveridge (1957), Pitman (1974); ZAIRE: Laurent (1956) (1960) (1965), Witte (1962); ZAMBIA: Broadley (1971e) (1973), Vesey-Fitzgerald (1958), Wilson (1965); ZIMBABWE: Broadley (1959), Broadley and Cock (1975); Fitzsimons (1943) (1974).

Appendix 3

CROCODILIANS AND CHELONIANS

Order: Crocodilia
Family: Crocodilidae
Reference: Abercrombie (1978); Baecker (1940); Beck (1978); Boulenger
(1889a); Cott (1961); Diefenbach (1973); Gabe (1971) (1972);
Guggisberg (1972); Hara and Kikuchi (1978); Jacobshagen
(1920); Pooley (1962) (1969); Schmidt and Inger (1957); Sill
(1968); Sims and Singh (1978); Teichner (1978); Villiers (1958);
Vogel (1964); Wermuth (1953); Wermuth and Fuchs (1978).

Genus: *Crocodylus* Laurenti
Crocodylus cataphractus cataphractus Cuvier
Crocodylus cataphractus congicus Fuchs, Mertens and Wermuth
Crocodylus niloticus niloticus Laurenti
Crocodylus niloticus africanus Laurenti
Cordodylus niloticus chamses Bory
Crocodylus niloticus cowiei (Smith)
Crocodylus niloticus madagascariensis Grandidier
Crocodylus niloticus pauciscutatus Deraniyagala
Crocodylus niloticus suchus Geoffroy

Genus: *Osteolaemus* Cope
Osteolaemus tetraspis tetraspis Cope
Osteolaemus tetraspis osborni (Schmidt)

Order: Chelonia

Reference: Arnold and Burton (1978); Baldwin and Lofton (1959); Bickham (1975); Bons and Bonaric (1971); Boulenger (1889a); Brongersma (1967); Burchfield (1975); Bustard (1971) (1972); Bustard and Greenham (1968); Caldwell and others (1959); Carr and Ogren (1959); Cloudsley-Thompson (1970); Frair (1979); Frair and Prol (1978); Gans and Hughes (1967); Gaymer (1968); Haacke (1965); Hendrikson (1958); Hirth (1971); Honegger (1967); Hughes and others (1968); Issekutz and Vegh (1928); Khalil and Haggar (1955); Kramer and Richardson (1979); Loveridge (1941e); Loveridge and Williams (1957); McDowell (1964); Merkle (1975); Peters (1969); Phillips (1977); Porter (1972); Pritchard (1967) (1971); Raj (1976); Raw (1978b); Schmidt and Inger (1957); Spoczynska (1972); Urban (1970); Vasilescu (1970); Villiers (1958); Vogel (1964); Wermuth (1977)

Suborder: Pleurodira
Family: Pelomedusidae

Genus: *Erymnochelys* Baur
Erymnochelys madagascariensis (Grandidier)

Genus: *Pelomedusa* Wagler
Pelomedusa subrufa subrufa (Lacepede)
Pelomedusa subrufa olivacea (Schweigger)

Genus: *Pelusios* Wagler
Pelusios adansonii (Schweigger)
Pelusios bechuanicus Fitzsimons
Pelusios carinatus Laurent
Pelusios castaneus (Schweigger)
Pelusios gabonensis (Dumeril)
Pelusios nanus Laurent
Pelusios niger (Dumeril and Bibron)
Pelusios rhodesianus Hewitt
Pelusios sinuatus (Smith)
Pelusios subniger (Lacepede)

Suborder: Cryptodira
Superfamily: Testudinioidea
Family: Testudinidae

Genus: *Acinixys* Siebenrock
Acinixys planicauda (Grandidier)

Genus: *Chersina* Gray
Chersina angulata Gray

Genus: *Geochelone* Fitzinger
Geochelone (Aldabrachelys) gigantea gigantea (Schweigger)
Geochelone (Aldabrachelys) gigantea daudinii (Dumeril and Bibron)
Geochelone (Asterochelys) radiata (Shaw)
Geochelone (Asterochelys) yniphora (Vaillant)
Geochelone (Geochelone) pardalis pardalis (Bell)
Geochelone (Geochelone) pardalis babcocki (Loveridge)
Geochelone (Geochelone) sulcata (Gmelin)

Genus: *Homopus* Dumeril and Bibron
Homopus aereolatus (Thumberg)
Homopus boulengeri Duerden
Homopus femoralis Boulenger
Homopus signatus (Walbaum)

Genus: *Kinixys* Bell
Kinixys belliana belliana Gray
Kinixys belliana nogueyi (Lataste)
Kinixys erosa (Schweigger)
Kinixys homeana Bell

Genus: *Malacochersus* Lindholm
Malacochersus tornieri (Siebenrock)

Genus: *Psammobates* Fitzinger
Psammobates geometricus (Linnaeus)
Psammobates oculifer (Kuhl)

Psammobates tentorius tentorius (Bell)
Psammobates tentorius trimeni (Boulenger)
Psammobates tentorius verroxii (Smith)

Genus: *Pyxis* Bell
Pyxis arachnoides Bell

Genus: *Testudo* Linnaeus
Testudo (Testudo) graeca graeca Linnaeus
Testudo (Testudo) graeca terrestris Forskal
Testudo (Pseudotestudo) kleinmanni Lortet

Family: Emydidae
Subfamily: Emydinae

Genus: *Emys* Dumeril
Emys orbicularis (Linnaeus)

Family: Emydidae
Subfamily: Batagurinae

Genus: *Mauremys* Gray
Mauremys leprosa (Schweigger)

Superfamily: Chelonioidea
Family: Cheloniidae
Subfamily: Cheloniinae
Tribe: Cheloniini

Genus: *Chelonia* Latreille
Chelonia mydas (Linnaeus)

Family: Cheloniidae
Subfamily: Cheloniinae
Tribe: Carettini

Genus: *Caretta* Merrem
Caretta caretta (Linnaeus)

Genus: *Eretmochelys* Fitzinger
Eretmochelys imbricata (Linnaeus)

Genus: *Lepidochelys* Fitzinger
Lepidochelys olivacea (Eschscholtz)

Family: Cheloniidae
Subfamily: Dermochelyinae

Genus: *Dermochelys* Blainville
Dermochelys coriacea (Linnaeus)

Superfamily: Trionychoidea
Family: Trionychidae

Genus: *Cyclanorbis* Gray
Cyclanorbis elegans (Gray)
Cyclanorbis senegalensis (Dumeril and Bibron)

Genus: *Cycloderma* Peters
Cycloderma aubryi (Dumeril)
Cycloderma frenatum Peters

Genus: *Trionyx* Geoffroy
Trionyx triunguis (Forskal)

BIBLIOGRAPHY

Abercrombie, Clarence L. (1978). Notes on West African Crocodilians. *J. Herpet. 12*, 260-262.

Aellen, V. (1951). Contribution a l'herpetologie du Maroc. *Bull. Soc. Sci. Nat. Phys. Maroc. Rabat 31*, 153-199.

Akester, John. (1979). Male combat in captive Gaboon Vipers (Serpentes: Viperidae). *Herpetologica 35*, 124-128

Algauhari, A. E. J. (1967). Experimental studies on the blood sugar in reptiles. Effect of fasting, temperature and insulin administration in *Psammophis sibilans*. *Z. vergl. Physiol. 54*, 395-399.

Anderson, John. (1892). On a small collection of mammals, reptiles, and batrachians from Barbary. *Proc. Zool. soc. London* 3-24.

Anderson, John. (1893). On a new species of *Zamenis* and a new species of *Bufo* from Egypt. *Ann. Mag. Nat. Hist. (ser. 6) 12*, 439.

Anderson, John (1896). *A contribution to the herpetology of Arabia with a preliminary list of the Reptiles and Amphibians of Egypt*. London, pp. 122.

Anderson, John (1898). *Zoology of Egypt. I. Reptilia and Batrachia*. Bernard Quaritch, London, pp. 371.

Anderson, John (1900). On the presence of *Echis coloratus* Gunther in Africa. *Ann. Mag. Nat. Hist. (ser. 7) 6*, 419-425.

Andersson, Lars G. (1901). Some new snakes from Cameroon and South America, belonging to the collections of the Royal Museum in Stockholm. *Bih. K. Svenska Vet.-Akud. Hand. 27*, 1-26.

Andersson, Lars G. (1903). Neue Batrachier aus Kamerun von den Herren Dr. Y. SJOSTEDT and Dr. S. JUNGER gesammelt. *Verh. zool. bot. Ges. Wien 53*, 141-145.

Andersson, Lars G. (1908). A remarkable new gecko from South Africa in the Natural History Museum Wiesbaden. *Jahrb. Ver. Nat. Wiesbaden 61*, 299-300.

Andersson, Lars G. (1910). A new species of *Hemidactylus* from Harrar, Abyssinia. *Jahrb. Ver. Nat. Wiesbaden 63*, 200-205.

Andersson, Lars G. (1912). Notes on *Hemidactylus tropidolepis* Mocq., *Jahrb. Ver. Nat. Wiesbaden 65*, 227-231.

Andersson, Lars G. (1935). Reptiles and Batrachians from the Central Sahara. *Goteborg Kungl. Vetensk. Vitterh. Samh. Handl. B4, 10*, 4-10.

Andersson, Lars G. (1937). Reptiles and Batrachians collected in the Gambia by Gustav Svensson and Birger Rudebeck (Swedish Expedition 1931). *Arkiv. Zool., A29 (16)*, 1-28.

Andren, Claes (1976). The reptile fauna in the lower alpine zone of Aberdare and Mount Kenya. *Brit. J. Herpet. 5*, 566-575.

Angel, Fernand (1920). Liste de reptiles du Haut-Zambeze et de l'Afrique australe. Description d'une espece nouvelle de genre *Monopeltis*. *Bull. Mus. Hist. nat. Paris 26*, 614-617.

Angel, Fernand (1921). Contribution a l'etude des chameleons de Madagascar. *Bull. Mus. Hist. nat. Paris 27*, 328-331 and 406-412.

Angel, Fernand (1922a). Sur une collection de Reptiles et de Batraciens, recueillis au Soudan francais par le Mission du Dr. Millet Horsin. *Bull. Mus. Hist. nat. Paris 28*, 39-41.

Angel, Fernand (1922b). Sur un lezard d'un Genre nouveau de la Famille des Gerrhosauridae. *Bull. Mus. Hist. nat. Paris 28*, 150-152.

Angel, Fernand (1924a). Sur une forme nouvelle de lezard, a provenance de Madagascar appartenant au genre *Grandidierina*. *Bull. Mus. Hist. nat. Paris 30*, 450-452.

Angel, Fernand (1924b). Reptiles. *in Mission Roban-Chabot. Angola et Rhodesia* (1912-1914). Paris (1923), *4*, 157-169.

Angel, Fernand (1929a). Description d'un gecko nouveau de Madagascar. *Bull. soc. zool. France 54*, 489.

Angel, Fernand (1929b). Contribution a l'etude systematique des Lezards appartenant aux genres *Uroplatus* et *Brookesia*. *Mem. Acad. Malagache 9*, 1-63.

Angel, Fernand (1930a). Reptiles et Batraciens recueillis par M. MONARD au Cameroun. *Bull. Mus. Hist. nat. Paris 2*, 253-254.

Angel, Fernand (1930b). Sur un lezard nouveau de Madagascar, appartenant au genre *Lygodactylus*. *Bull. soc. zool. France 55*, 253.

Angel, Fernand (1930c). Diagnoses d'especes nouvelles de lezards de Madagascar, appartenant au genre *Scelotes*. *Bull. Mus. Hist. nat. Paris (ser. 2) 2*, 506-509.

Angel, Fernand (1930d). Sur *Uroplatus phantasticus* Boulgr. et *Uroplatus schneideri* Lamberton. *Bull. Acad. Malagache (new ser.) 13*, 105-108.

Angel, Fernand (1930e). Sur l'hahitat d'un certain nombre de Lezards de Madagascar. *Bull. Acad. Malagache (new ser.) 13*, 109-116.

Angel, Fernand (1933a). Lezards nouveaux de Madagascar, appartenant au genre *Scelotes*. *Bull. soc. zool. France 58*, 294-195.

Angel, Fernand (1933b). Sur quelques Reptiles et Batraciens du Nord du Soudan francais. *Bull. Mus. Hist. nat. Paris (ser. 2) 5*, 68-69.

Angel, Fernand (1933c). *Les serpents de l'Afrique Occidentale Francais*. Larose edition, Paris, pp. 246.

Angel, Fernand (1933d). Sur un genre malagache nouveau de la famille des Chamaeleontides. *Bull. Mus. Hist. nat. Paris (ser. 2) 6*, 443-446.

Angel, Fernand (1934). Remarques sur le genre *Oophilositum* Parker (Colubridae, Aglyphe) et description d'une espece nouvelle. *Bull. soc. zool. France 59*, 417-419.

Angel, Fernand (1935). Lezards des Iles du Cap Vert, rapportes par le Professeur Chevalier. Description de deux especes nouvelles. *Bull. Mus. Hist. nat. Paris (ser. 2) 7*, 165-169.

Angel, Fernand (1936a). Description d'une nouvelle espece du genre *Philochortus* provenant des confins saharo-soudanais et tableau synoptique des especes du genre. *Bull. soc. zool. France 61*, 100-105.

Angel, Fernand (1936b). Sur quelques formes nouvelles de Reptiles et de Batraciens du Sahara central. *Bull. soc. zool. France 61*, 273-277.

Angel, Fernand (1936c). Deux Gekkos nouveaux de Madagascar appartenant au genre *Phyllodactylus*. *Bull. soc. zool. France 61*, 508.

Angel, Fernand (1936d). Materiaux erpetologiques recueillis a Madagascar par R. Heim. *Bull. Mus. Hist. nat. Paris (ser. 2) 8*, 125-139.

Angel, Fernand (1938a). Liste des reptiles de Mauritanie recueillis par la mission d'etudes de la biologie des Acridiens en 1936 et 1937. Description d'une sous-especie nouvelle d'*Eryx muelleri*. *Bull. Mus. Hist. nat. Paris (ser. 2) 10*, 485-487.

Angel, Fernand (1938b). Lezards Scincide et Chamaeleontide nouveaux de Madagascar. *Bull. Mus. Hist. nat. (ser. 2) 10,* 574-577.

Angel, Fernand (1940). Reptiles et Amphibiens recueillis au Cameroun par la mission, P. LEPESME, R. PAULIAN et A. VILLIERS (3e note). *Bull. Mus. Hist. nat. Paris (ser. 2) 12,* 390-399.

Angel, Fernand (1942). Les Lezards de Madagascar. *Mem. Acad. Madagascar 36,* 1-193.

Angel, Fernand (1943). Sur deux lezards nouveaux de la Haute-Guinee francaise, appartenant aux familles de amphisbenides et des gekkonides, Materiaux de la Mission Lamotte au Mont Nimba en 1942. Premiere note. *Bull. Mus. Hist. nat. Paris (ser. 2) 15,* 163-166.

Angel, Fernand (1944). Contribution a l'etude de la faune herpetologique du Sahara central. *Bull. Mus. Hist. nat. Paris (ser. 2) 16,* 418-419.

Angel, Fernand (1949a). Sur une collection de Reptiles de Madagascar rapportes au Museum par M. le professeur Millot. Description d'une espece nouvelle du genre *Paracontias. Mem. Inst. sci. Madagascar A3,* 81-87.

Angel, Fernand (1949b). Deuxieme note sur les reptiles de Madagascar recoltes par le Professeur Millot. Description d'un lezard scincide nouveau du genre *Scelotes (S. trilineatus* sp. nov.). *Mem. Inst. sci. Madagascar A3,* 157-164.

Angel, Fernand, Guibe, Jean and Maxime Lamotte (1954). La reserve naturelle integrale du Mont Nimba. 21. Lezards. *Mem. I. F. A. N. 40,* 371-379.

Angel, Fernand and Lhote, H. (1938). Reptiles et Amphibiens du Sahara central et du Soudan. *Bull. Com. Et. hist. sci. Afr. occ. franc. 21,* 345-384.

Anthony, J. and Guibe, J. (1951a). *Casarea,* forme de passage entre les Boides et les serpents proteroglyphes. *C. r. Acad. Sci. Paris, 233,* 203-204.

Anthony, J. and Guibe J. (1951b). Les affinites anatomiques de *Bolyeria* et de *Casarea* (Ophidiens). *Mem. Inst. sci. Madagascar A7,* 189-201.

Arnold, E. N. (1973). Relationships of the Palaearctic lizards assigned to the genera *Lacerta, Algyroides* and *Psammodromus* (Reptilia: Lacertidae). *Bull. Brit. Mus. nat. Hist. (Zool.) 25,* 289-366.

Arnold, E. N. and Burton, J. A. (1978). *A field guide to the reptiles and amphibians of Britain and Europe.* Collins, London pp. 272.

Arnold, E. N. and Leviton A. E. (1977). A revision of the lizard genus *Scincus* (Reptilia: Scincidae). *Bull. Brit. Mus. nat. Hist. (Zool.) 31,* 187-248.

Aruo, S. K. (1977). Parasites of the common African Python, *Python sebae. E. Afr. Wildlife J. 15,* 159-164.

Atsatt, S. R. (1953). Storage of sperm in the female chamaeleon, *Microsaura pumila pumila. Copeia,* 59.

Audouin, Victor J. V. (1827). Explication sommaire des Planches de Reptiles. In, Savigny, M. J. C. L. de, *Description, Egypte, Histoire naturelle Reptiles.* Paris, 1, 161-184.

Auffenberg, W. (1978). Courtship and breeding behavior in *Geochelone radiata* (Testudines: Testudinidae). *Herpetologica 34,* 277-287.

Avery, D. F. and Tanner, W. W. (1971). Evolution of the iguanine lizards (Sauria, Iguanidae) as determined by osteological and myological characters. *Brigham Young Univ. Sci. Bull. Biol. 12,* 1-79.

Aylmer, G. (1922). The snakes of Sierra Leone. *Sierra Leone Studies 5,* 7-37.

Badir N. (1959). Seasonal variation of the male urogenital organs of *Scincus scincus* L. and *Chalcides ocellatus* Forsk. *Zeits. Wiss. Zool. 160,* 290-343.

Badir, N. (1968a). Structure and function of corpus luteum during gestation in the viviparous lizard *Chalcides ocellatus. Anat. Anz. 122,* 1-10.

Badir, N. (1968b). The effect of population density on the embryonic mortality in the viviparous lizard *Chalcides ocellatus* (Forsk). *Anat. Anz. 122,* 11-14.

Badir, N. and Hussien, M. F. (1965). Effect of temperature, food and illumination on the reproduction of *Chalcides ocellatus* (Forsk) and *Scincus scincus* (Linn.). *Bull. Fac. Sci. Cairo Univ. 39*, 179-185.

Baecker, R. (1940). Uber die als Stratum fibrosum (compactum) bezeichnete Grenzschicht im Verdauungs-Kanal der Wirbeltiere. *Z. Forsch. mikrosk. Anat. 47*, 49-99.

Baldwin, William P. Jr. and Lofton, John P. Jr. (1959). The Atlantic loggerhead sea turtle, *Caretta caretta caretta* (L.) in America. III. The loggerhead turtles of Cape Romain, South Carolina. *Bull. Florida St. Mus. 4*, 319-348.

Balozet, L. (1957). La vipere lebetine et son venin. *Arch. Inst. Pasteur Algerie 35*, 220-295.

Banks, B. E. C., Miledi, R. and R. A. Shipolini (1974). The primay sequences and neuromuscular effects of three neurotoxic polypeptides from the venom of *Dendroaspis viridis*. *Eur. J. Biochem. 45*, 457-468.

Bannister, L. H. (1968). Fine structure of the sensory endings in the vomero-nasal organ of the slow-worm *Anguis fragilis*. *Nature* (London) *217*, 275-276.

Barbault, R. (1967). Recherches ecologiques dans la savane de Lamto (Cote d'Ivoire): Le cycle annuel de la biomasse des amphibiens et des lezards. *La Terre et la Vie 21*, 297-318.

Barbault, R. (1971a). Recherches ecologiques dans la savane de Lamto (Cote d'Ivoire). Production annuelle des populations naturelles du lezard *Mabuya buttneri* (Matschie). *La Terre et la Vie 25*, 203-217.

Barbault, R. (1971b). Les peuplements d'Ophidiens des savanes de Lamto (Cote d'Ivoire). *Ann. Univ. Adidjan* (ser. E.) *4*, 133-194.

Barbault, R. (1974a). Ecologie comparee des lezards *Mabuya blandingii* (Hallowell) et *Panaspis kitsoni* (Boulenger) dans les forets de Lamto (Cote d'Ivoire). *La Terre et al Vie 28*, 95-117.

Barbault, R. (1974b). Dynamique des populations naturelles du lezard *Mabuya maculilabris* dans les savanes de Lamto (Cote d'Ivoire). *Bioecos 1*.

Barbault, R. (1974c). Observations ecologiques dans la savane de Lamto (Cote d'Ivoire): structure trophique de l'herpetocenose. *Bull. Ecol. 5*, 7-25.

Barbault, R. (1974d). Structure et dynamique des populations naturelles du lezard *Mabuya buettneri* dans les savanes de Lamto (Cote d'Ivoire). *Bull. Ecol. 5*, 105-121.

Barbault, R. (1974e). Structure et dynamique d'un peuplement de Lezards: les scincides de la savane de Lamto (Cote d'Ivoire). *La Terre et la Vie 28*, 352-428.

Barbault, R. (1974f). Dynamique des populations naturelles du lezard *Panaspis nimbaensis* dans les savanes de Lamto (Cote d'Ivoire). *Bull. soc. zool. France 99*, 345-361.

Barbault, R. (1975). Place des Lezards dans la biocenose de Lamto: relations Trophiques; production et consommation des populations naturelles. *Bull. I. F. A. N. 37A*, 467-514.

Barbault, R. (1976). Population dynamics and reproductive patterns of three African skinks. *Copeia* 483-490.

Barbour, T. (1911). A new race of chameleon from British East Africa. *Proc. Biol. Soc. Washington 24*, 219-220.

Barbour, T. (1914). Some new reptiles. *Proc. New England zool. Club 4*, 95-98.

Barbour, T. (1918). Vertebrata from Madagascar. Amphibians and Reptiles. *Bull. Mus. Comp. Zool. 61*, 479-489.

Barbour, T. and Loveridge A. (1928a). A comparative study of the herpetological fauna of the Uluguru and Usambara Mountains, Tanganyika Territory, with descriptions of new species. *Mem. Mus. Comp. Zool. Harvard 50*, 87-265.

Barbour, T. and Loveridge A. (1928b). New skinks of the genus *Scelotes* from Mozambique and Madagascar. *Proc. New England zool. Club 10,* 63-65.

Barry T. H. (1953). Contributions to the cranial morphology of *Agama hispida* (Linn.). *Ann. Univ. Stellenbosch (A) 29,* 55-77.

Barzilay, M., Kaminsky, E. and Condrea, E. (1978). Exposure of human red blood cell membrance phospholipids to snake venom phospholipases A — II. Hydrolysis of substrates in intact and resealed cells by phospholipase from Ringhals (*Hemachatus haemachatus*) venom: effect of calcium ions. *Toxicon 16,* 153-161.

Battersby, J. C. (1950). A new amphisbaenid lizard from Tanganyika Territory and notes on the rare snake *Chilorhinophis. Ann. Mag. Nat. Hist. (ser. 12) 3,* 413-417.

Beck, C. (1978). Breeding the West African dwarf crocodile *Osteolaemus tetraspis tetraspis* at Memphis Zoo. *Inter. Zoo Yrbk. 18,* 89-91.

Bedriaga, J. von. (1884). Die neue Lacertiden-Gattung *Latastia* und ihre Arten. *Ann. Mus. Genova 20,* 307-324.

Behler, J. L. and Brazaitis, P. (1974). Breeding the Egyptian cobra, *Naja haje,* at the New York Zoological Park. *Inter. Zoo Yrbk. 14,* 83-84.

Beiran, D. and Currie, D. (1967). Snake bite due to *Thelotornis kirtlandi. Centr. Afr. J. Med. 13,* 137-139.

Bellairs, A.d'A. and Shute, C. C. D. (1954). Notes on the herpetology of an Algerian beach. *Copeia* 224-226.

Bengis, R. G. and Noble, D. F. (1976). Postsynaptic blockage of neuromuscular transmission by toxin II from the venom of the South African ringhals cobra (*Hemachatus haemachatus*). *Toxicon 14,* 167-173.

Bentley, Peter J. (1976). Osmoregulation, in, *Biology of the Reptilia,* editors: Gans, C. and Dawson, W. R. *5,* 365-412.

Bhattacharya, S. and Gaitonde, B. B. (1979). Partial purification of cholinesterase from the venom of the saw scaled viper (*Echis carinatus*). *Toxicon 17,* 429-431.

Bianconi, J. J. (1849). Specimina zoologica Mossambicana (De piscibus et reptilibus). *Nuovi Ann. Sci. Nat. Bologna (ser. 2) 10,* 106-109 and 202-208.

Bickham, John W. (1975). A cytosystematic study of turtles in the genera *Clemmys, Mauremys* and *Sacalia. Herpetologica 31,* 198-204.

Blanc, Charles P. (1965). Etudes sur les Iguanidae de Madagascar. I. Le squelette de *Chalarodon madagascariensis* Peters 1854. *Mem. Mus. natl. Hist. nat. Paris (N. S.) A. Zool. 33,* 93-146.

Blanc, Charles P. (1966). Notes sur les Gekkonidae de Madagascar. I. Observations relatives a deux especes de *Geckolepis: G. maculata* Peters 1880 et *G. typica* Grandidier, 1867. *Vie et Milieu 17,* 453-460.

Blanc, Charles P. (1967). Notes sur les Gerrhosaurinae de Madagascar. I. Observations sur *Zonosaurus maximus* Boulenger 1896. *Ann. Fac. Sci. Univ. Madagascar 5,* 107-116.

Blanc, Charles P. (1969). Etudes sur les Iguanidae de Madagascar. II. Observations sur l'ecologie de *Chalarodon madagascariensis* Peters. *Oecolgie 2,* 292-318.

Blanc, Charles P. (1971). Les reptiles de Madagascar et des iles viosines. *Ann. Univ. Madagascar 8,* 95-178.

Blanc, Charles P. and Blanc F. (1967). Observations biologiques (adultes, oeufs, jeunes) sur quelques sauriens du Mont Bity. *Ann. Fac. Sci. Univ. Madagascar 5,* 67-74.

Blanc, Charles P. and Carpenter C. C. (1968). Studies on the Iguanidae of Madagascar. III. Social and reproductive behavior of *Chalarodon madagascariensis. J. Herpet. 3,* 125-134.

Blanc, M. (1936). *Fauna Tunisienne, Reptiles et Batraciens.* Tunis.

Blanford, William T. (1870). *Observations on the Geology and Zoology of Abyssinia.* London.

Blanford, William T. (1874). Description of new lizards from Persia and Baluchistan. *Ann. Mag. Nat. Hist. (ser. 4) 13,* 453-455.

Blyth, Edward. (1861). Report of the Curator, Zoological Department. *J. Asiatic Soc. Bengal 1861, 29,* 87-115.

Bocage J. V. du B. (1866a). Lista dos reptis das possessoes portuguezas d'Africa occidental que existem no Museu Lisboa. *J. Sci. Lisbon 1,* 37-56.

Bocage J. V. du B. (1866b). Reptiles nouveaux ou peu connus recueilles dans les possessions portugaises de l'Afrique occidentale, qui se trouvent au meseum de Lisbonne. *J. Sci. Lisbon 1,* 57-78.

Bocage J. V. du B. (1867a). Diagnoses des quelques reptiles nouveaux le l'Afrique occidentale. *J. Sci. Lisbon 1,* 229-232.

Bocage, J. V. du B. (1867b). Descriptions of two new Saurians from Mossamedes (West Africa) *Ann. Mag. Nat. Hist. (ser. 3) 20,* 225-228.

Bocage J. V. du B. (1870). Description d'un Saurien nouveau de l'Afrique occidntale. *J. Sci. Lisbon 3,* 66-68.

Bocage, J. V. du B. (1872). Diagnoses de quelques especes nouvelles de reptiles d'Afrique occidentale. *J. Acad. Sci. Lisbon 4,* 72-82.

Bocage, J. V. du B. (1873a). Melanges erpetologiques. II. Sur quelques reptiles et batraciens nouveaux, rares ou peu connus d'Afrique occidentale. *J. Acad. Sci. Lisbon 4,* 209-227.

Bocage, J. V. du B. (1873b). Reptiles nouveaux de l'interieur de Mossamedes. *J. Acad. Sci. Lisbon 4,* 247-253.

Bocage, J. V. du B. (1875). Sur deux reptiles nouveaux de l'Archipel de Cap-Vert. *J. Acad. Sci. Lisbon 5,* 108-112.

Bocage, J. V. du B. (1879). Some new reptiles and batrachians from Angola. *J. Acad. Sci. Lisbon 7,* 87-99.

Bocage, J. V. du B. (1882a). Reptiles rares ou nouveaux d'Angola. *J. Acad. Sci. Lisbon 8,* 299-304.

Bocage, J. V. du B. (1886a). Reptis e Amphibios de S. Thome. *J. Acad. Sci. Lisbon 11,* 65-70.

Bocage, J. V. du B. (1886b). Reptiles et Batraciens nouveaux de l'Ile de St. Thome. *J. Acad. Sci. Lisbon 11,* 71-75.

Bocage, J. V. du B. (1886c). Note additionnelle sur les Reptiles de St. Thome. *J. Acad. Sci. Lisbon 11,* 103-104.

Bocage, J. V. du B. (1886e). Typhlopiens nouveaux de la Faune africaine. *J. Acad. Sci. Lisbon 11,* 171-174.

Bocage, J. V. du B. (1887). Zoologia Melanges erpetologiques. V. Reptiles et Batraciens de Quissange, Benguella, envoyes par M. J. d'Anchieta. *J. Acad. Sci. Lisbon 11,* 177-211.

Bocage, J. V. du B. (1889a). Melanges herpetologiques. *J. Acad. Sci. Lisbon (ser. 2) 1,* 125.

Bocage, J. V. du B. (1893a). Mammiferes, Aves e Reptis de Ilha de Anno-Bom. *J. Acad. Sci. Lisbon (ser. 2) 3,* 45-46.

Bocage, J. V. du B. (1893b). Diagnoses de quelques nouvelles especes de reptiles et batraciens d'Angola. *J. Acad. Sci. Lisbon (ser. 2) 3,* 115-121.

Bocage, J. V. du B. (1895a). *Herpetologie d'Angola et du Congo.* Lisbon, Imp. Natl., pp. 203.

Bocage, J. V. du B. (1896). Sur deux Agames d'Angola a ecaillure heterogene. *J. Acad. Sci. Lisbon (ser. 2) 4,* 127-130.

Bocourt, Firmin (1870). Description de quelques Sauriens nouveaux originaires de L'Amerique meridionale. *Arch. Mus. Hist. Nat. Paris (ser. 2) 6,* 11-18.

Bocourt, Firmin (1875). Note sur une nouvelle espece d'Ophidien. *Ann. Sci. nat. Zool. 2, 3.*

Boettger, C. R. and Muller, L. (1914). Preliminary notes on the local races of some Canarian lizards. *Ann. Mag. Nat. Hist. (ser. 8) 14,* 67-78.

Boettger, Oskar (1874). Reptilien von Marocco und den canarischen Inseln. *Abhandl. Senckenb. naturf. Gess. 9,* 121-192.

Boettger, Oskar (1877). Die reptilien und amphibien von Madagascar. *Abhandl. Senckenb. naturf. Gess. 11,* 1-56.

Boettger, Oskar (1880a). Diagnoses reptilium et batrachoirum novorum a Carlo Ebenau in insula Nossi-Be Madagascariensi. *Zool. Anz. 4,* 279-283.

Boettger, Oskar (1880b). Diagnoses reptilium novorum Maroccanorum. *Zool. Anz. 4,* 570-572.

Boettger, Oskar (1880c). Die Reptilien und Amphibien von Syrien, Palestina und Cypren. *Abhandl. Senckenb. naturf. Gess.* 132-219.

Boettger, Oskar (1883). Die reptilien und amphibien von Marocco. II. *Abhandl. Senckenb. naturf. Gess. 13,* 93-146.

Boettger, Oskar (1886). Beitrage zur Herpetologie und Malakozoologie Sudwest-Afrikas. I. Zur Kenntnis der Fauna von Angra Pequenia. *Abhandl. Senckenberg naturf. Gess.* 3-29.

Boettger, Oskar (1887). Zweiter Beitrag zur Herpetologie Sudwest- und Sud-Afrikas. *Abhandl. Senckenb. naturf. Gess.* 135-173.

Boettger, Oskar (1888). Zweiter Beitrag zur Herpetologie Sudwest- und Sud-Afrikas. Batrachier. *Abhandl. Senckenb. Naturf. Gess.* 13-108.

Boettger, Oskar (1889). Herpetologische Miscellen. V. Transvaal; VI. Pondoland. *Abhandl. Senckenb. naturf. Gess.* 286-295.

Boettger, Oskar (1892b). Drei neue colubriforme Schlangen. *Zool. Anz. 15,* 417-420.

Boettger, Oskar (1893a). *Katalog der Reptilien-Sammlung in Museum der Senckenbergischen Naturforschenden Gesellschaft in Frankfurt am Main. I. Teil, Rhynchocephalen, Schidkroten, Krokodile, Eidechsen, Chamaleons.* Frankfurt, pp. 140.

Boettger, Oskar (1893b). Ubersicht der von Prof. C. Keller anlasslich der Ruspoli'schen Expedition nach den Somalilandern gesammelten. Reptilien und Batrachier. *Zool. Anz. 16,* 113-119 and 129-132.

Boettger, Oskar (1894a). Diagnosen eines Gecko und cines Chamaeleons aus Sud-Madagascar. *Zool. Anz. 17,* 137-140.

Boettger, Oskar (1894b). Eine neue *Brookesia* (Chamaeleontidae) aus Nossi Be. *Zool. Anz. 17,* 182-185.

Boettger, Oskar (1894c). Eine neue Eidechse aus Sudwest-Afrika. *Abhandl. Mus. Dresden 5,* 1.

Boettger, Oskar (1895). Zwei neue Reptilien vom Zambesi. *Zool. Anz. 18,* 62-63.

Boettger, Oskar (1898). *Katalog der Reptilien-Sammlung in Museum der Senckenbergischen Naturforschenden Gesellschaft in Frankfurt am Main. II. Schlangen.* Frankfurt, pp. 160.

Boettger, Oskar (1913). Reptilien und Amphibien von Madagascar, den inseln und dem Festland Ostafrikas. *Reise in Ostafrika von Prof. A. Voeltzkow,* Band III, Syst. Arb. Heft 4.

Bogert, C. M. (1940). Herpetological results of the Vernay Angola Expedition with notes on African reptiles in other collections. Part I. Snakes, including an arrangement of African Colubridae. *Bull. Amer. Mus. Nat. Hist. 77,* 1-107.

Boie, F. (1826). Generalubersicht der Familien und Gattungen der Ophidier. *Isis von Oken (Jena) 1826,* 981-982.

Boie, F. (1827). Bemerkungen uber Merremis Versuch eines Systems der Amphibiens. 1te. Lieferung: Ophidier. *Isis von Oken (Jena) 20,* 508-566.

Bolanos, R., Flores, A., Taylor, R. T. and Cerdas, L. (1974). Color patterns and venom characteristics in *Pelamis platurus. Copeia* 909-911.

Boltt, R. E. and Ewer, R. F. (1964). The functional anatomy of the head of the puff adder, *Bitis arietans* (Merr.). *J. Morph. 114*, 83-106.

Bonin, James J. (1965). The eye of *Agamodon anguliceps* Peters (Reptilia, Amphisbaenia). *Copeia* 324-331.

Bons, J. (1957). Amphibiens et Reptiles recoltes dans le Sud-Ouest marocain. *C. r. Soc. Sc. nat. phys. Maroc. 23*, 96-98.

Bons, J. (1958). Contribution a l'etude de l'Herpetofaune marocaine. Reptiles de la region d'Ifrane. *Bull. Soc. Sc. nat. phys. Maroc. 38*, 167-182.

Bons, J. (1959). Les Lacertiliens de Sud-Ouest marocain. *Trav. Inst. sci. cherif, Zool. 18*, 1-130.

Bons, J. (1960a). Description d'un nouveau lezard du Sahara: *Eremias pasteuri* sp. nov. (Lacertides). *C. r. Soc. Sci. nat. phys. Maroc. 4*, 69-71.

Bons, J. (1960b). Effets de l'amincissement de la roque de l'oeuf sur le developpement de lezard *Acanthodactylus pardalis*. *C. r. Soc. Biol. Paris 154*, 490-492.

Bons, J. (1962). Notes sur troix couleuvres Africaines: *Coluber algirus, Coluber florulentus* et *Coluber hippocrepis*. Description de *Coluber algirus villiersi* subsp. nov. *Bull. Soc. Sci. nat. phys. Maroc 42*, 61-86.

Bons J. (1963). Notes sur *Blanus cinereus* (Vandelli), description d'une sous-espece Marocaine: *Blanus cinereus mettetali* subsp. nov. *Bull. Soc. Sci. nat. phys. Maroc. 43*, 95-107.

Bons, J. (1964). Note sur la ponte du lezard *Agama bibroni* (Sauria, Agamidae). *Bull. Soc. Sci. nat. phys. Maroc 44*, 65-71.

Bons, J. (1968a). Addition a la faune du Maroc: *Acanthodactylus savignyi savignyi* Aud. (Sauria, Lacertidae). *Bull. Soc. Sci. nat. phys. Maroc 47*, 387-394.

Bons, J. (1968b). Comportement d'*Agama bibroni* A. Dumeril 1851 (Sauria) durant la periode de reproduction. *Bull. Soc. Sci. nat. phys. Maroc 48*, 93-99.

Bons, J. and Bons, N. (1959). Sur l'herpetofaune des Doukkala. *Bull. Soc. Sci. nat. phys. Maroc 39*, 117-128.

Bons, J. and Bons, N. (1960). Notes sur la reproduction et le developpement de *Chamaeleo chamaeleon* (L.) *Bull. Soc. Sci. nat. phys. Maroc 40*, 323-335.

Bons, J. and Bons, N. (1969) Un cas d'intersexualite complete chez un lezard anguide du Maroc, *Ophisaurus koellikeri* (Gunther 1873). *C. r. Hebd. Seanc. Abad. Sci. Paris 2680*, 695-696.

Bons, J. and Bonaric J. C. (1971). Description histologique de la "dent de l'eclosion" ou caroncule chez *Chelonia mydas* (L.) *Brit. J. Herpet. 4*, 202-206.

Bons, J. and Girot, B. (1963). Revision de l'espece *Acanthodactylus scutellatus* (Lacertidae, Saurien). *Bull. Soc. Sci. nat. phys. Maroc 42*, 311-334.

Bons, J. and Pasteur G. (1957a). Recentes captures de *Saurodactylus fasciatus* Werner et nouvelles observations sur le genre *Saurodactylus* (Gekkonides). *Bull. Soc. Sci. nat. phys. Maroc 37*, 57-65.

Bons, J. and Pasteur, G. (1957b). Nouvelles remarques sur les saurodactyles, avec description de *Saurodactylus mauritanicus brosseti* n. subsp. *Bull. Soc. Sci. nat. phys. Maroc 37*, 175-195.

Bons, N. (1962). Le cycle de ponte du reptile Lacertide: *Acanthodactylus erythrurus lineomaculatus*. *C. r. Acad. Sci. Paris 255*, 165-167.

Bons, N. (1963a). Le cycle d'activite sexuelle du male du lezard *Acanthodactylus erythrurus lineomaculatus*. *C. r. Acad. Sci. Paris 256*, 1021-1023.

Bons, N. (1963b). Action des injections d'hormones hypophysaires gonadotropes sur le tractus genital du Lacertide: *Acanthodactylus erythrurus lineomaculatus*. *C. r. Acad. Sci. Paris 257*, 1376-1378.

Bons, N. (1964). Degenerescence des follicules ovariens chez un lezard du Maroc: *Acanthodactylus erythrurus lineomaculatus*. *Bull. Soc. Sci. nat. phys. Marox 44*, 75-82.

Bons, N. (1967). Evolution des cellules interstitielles du testicule chez un Lacertide: *Acanthodactylus erythrurus lineomaculatus. Bull. Soc. Sci. nat. phys. Maroc 47,* 207-213.

Bons, N. (1969). Le cycle sexuel de male chez *Acanthodactylus erythrurus lineomaculatus* Dum. et Bibr. (Sauria, Lacertidae). *Bull. Soc. Sci. nat. phys. Maroc. 49,* 161-204.

Bons, N. (1971). Mise en evidence de la presence d'acide ribonucleique dans le noyau des spermatides agees d'*Acanthodactylus erythrurus lineomaculatus* Dum. et Bibr. (Lacertidae). *Z. Zellforsch. Mikrosk. Anat. 118,* 249-253.

Bons, N. (1972). Variations histophysiologiques de tractus genital femelle du lezard *Acanthodactylus erythrurus lineomaculatus* Dum. et Bibr. au cours du cycle annuel. *Bull. Soc. Sci. nat. phys. Maroc 52,* 59-120.

Bons, N. (1973). L'hypophyse d'*Acanthodactylus erythrurus lineomaculatus* (Lacertide du Maroc) et ses variations au cours du cycle sexuel. *Bull. Soc. Sci. nat. phys. Maroc 53,* 119-133.

Boquet, P., Izard, Y., Jouannet, M. and Meaume, J. (1966a) Recherches biochimiques et immunologiques sur le venin des serpents. I. Essais de seperation des antigenes du venin de *Naja nigricollis* par filtration sur Sephadex. *Annls. Inst. Pasteur Paris 111,* 719-732.

Boquet, P., Izard, Y., Jouannet, M. and Meaume, J. (1966b). Etude de deux antigenes toxiques du venin de *Naja nigricollis. C. r. Hebd. Seanc. Acad. Sci. Paris (ser D) 262,* 1134-1137.

Botes, D. P. (1972). Snake venom toxins. The amino acid sequences of toxins b and d from *Naja melanoleuca. J. biol. Chem. 247,* 2866-2871.

Botes, D. P., Strydom, D. J., Anderson, C. G. and Christensen, P. A. (1971). Snake venom toxins: Purification and properties of three toxins from *Naja nivea* (Cape Cobra) venom and the amino acid sequence of toxin δ. *J. biol. Chem. 246,* 3132.

Botes, D. P. and Viljoen, C. C. (1974a). Purification of phospholipase A from *Bitis gabonica* venom. *Toxicon 12,* 611.

Botes, D. P. and Viljoen, C. C. (1974b). *Bitis gabonica* venom. The amino acid sequence of phospholipase A. *J. biol. Chem. 249,* 3827.

Botes, D. P. and Viljoen, C. C. (1976). The amino acid sequence of three non-curarimimetic toxins from *Naja nivea* venom. *Biochim. biophys. Acta 446,* 1.

Boulenger, E. G. (1921). On some lizards of the genus *Chalcides. Proc. zool. soc. London,* 77-83.

Boulenger, G. A. (1878). Description d'un genre nouveau et d'une espece nouvelle de la famillie des amphisbenides. *Bull. Soc. Zool. France 3,* 300-303.

Boulenger, G. A. (1883b). Description of a new genus of Geckos. *Ann. Mag. Nat. Hist. (ser. 5) 11,* 174-176.

Boulenger, G. A. (1885a). *Catalogue of the lizards in the British Museum (Natural History). I.* London, pp. 436.

Boulenger, G. A. (1885b). *Catalogue of the lizards in the British Museum (Natural History). II.* London, pp. 497.

Boulenger, G. A. (1885c). Descriptions of three new species of Geckos. *Ann. Mag. Nat. Hist. (ser. 5) 16,* 474-475.

Boulenger, G. A. (1887a). *Catalogue of the lizards in the British Museum (Natural History). III.* London, pp. 575.

Boulenger, G. A. (1887b). Descriptions of new species of reptiles and batrachians in the British Museum (Nat. Hist.). *Ann. Mag. Nat. Hist. (ser. 5) 20,* 50-53.

Boulenger, G. A. (1888a). Description of two new chamaeleons from Nossi-Be, Madagascar. *Ann. Mag. Nat. Hist. (ser. 6) 1,* 22-23.

Boulenger, G. A. (1888b). Descriptions of new reptiles and batrachians from Madagascar, *Ann. Mag. Nat. Hist. (ser. 6) 1,* 101-103.

Boulenger, G. A. (1888c). On new or little known South African reptiles. *Ann Mag. Nat. Hist. (ser. 6) 2*, 136-141.

Boulenger, G. A. (1889a). *Catalogue of the Chelonians, Rhynchocephalians and Crocodiles in the British Museum (Natural History)*. London pp. 311.

Boulenger, G. A. (1889b). Descriptions of two new reptiles and batrachians from Madagascar, *Ann. Mag. Nat. Hist. (ser. 6) 4*, 244-248.

Boulenger, G. A. (1889c). Descriptions of new Typhlopidae in the British Museum. *Ann. Mag. Nat. Hist. (ser. 6) 4*, 360-363.

Boulenger, G. A. (1890a). Descriptions of a new snake of the genus *Glauconia* Gray, obtained by Dr. Emin Pasha on the Victoria Nyanza. *Ann. Mag. Nat. Hist. (ser. 6) 6*, 91-93.

Boulenger, G. A. (1890b). First report on additions to the lizard collection in the British Museum (Natural History). *Proc. zool. soc. London* 77-86.

Boulenger, G. A. (1891a). On some reptiles collected by Sig. L. Brichetti Robecchi in Somaliland. *Ann. Mus. Genova 2*, 5-15.

Boulenger, G. A. (1891b). Remarks on the genus *Heterolepis* Smith. *Ann. Mag. Nat. Hist. (ser. 6) 7*, 417.

Boulenger, G. A. (1892a). *Reptilia and Batrachia*. Appendix to Distants 'Naturalist in the Transvaal'. pp. 174-176.

Boulenger, G. A. (1892b). Description of a new snake from Nubia. *Ann. Mag. Nat. Hist. (ser. 6) 9*, 74-76.

Boulenger, G. A. (1893). *Catalogue of the snakes in the British Museum (Natural History)*. *I*. London, pp. 448.

Boulenger, G. A. (1894a). *Catalogue of the snakes in the British Museum (Natural History)*. *II*. London. pp. 382.

Boulenger, G. A. (1894b). On a new gecko from South Africa. *Proc. zool. soc. London* 608.

Boulenger, G. A. (1894e). Second report on additions to the lizard collection in the Natural History Museum. *Proc. zool. soc. London* 722-736.

Boulenger, G. A. (1895b). An account of the reptiles and batrachians collected by Dr. A. Donaldson Smith in western Somaliland and the Galla Country. *Proc. zool. soc. London* 530-540.

Boulenger, G. A. (1895d). On some new or little-known reptiles obtained by W. H. Crosse Esq. on the Niger. *Ann. Mag. Nat. Hist. (ser. 6) 16*, 32-34.

Boulenger, G. A. (1895g). On the reptiles and batrachians obtained by Mr. E. Lort-Phillips in Somaliland. *Ann. Mag. Nat. Hist. (ser. 6) 16*, 165-169.

Boulenger, G. A. (1895h). Descriptions of two new snakes from Usambara, German East Africa. *Ann. Mag. Nat. Hist. (ser. 6) 16*, 171-173.

Boulenger, G. A. (1895i). Rettili e Batraci. *in*, Esplorazione del Guiba e dei suoi Affluenti compeuta dal Cap. V. Bottego durante gli Anni 1892-93. *Ann. Mus. Civ. Stor. Nat. Genova (ser. 2) 15*, 9-18.

Boulenger, G. A. (1896a). *Catalogue of the snakes in the British Museum (Natural History)*. *III*. London, pp. 727,

Boulenger, G. A. (1896b). Descriptions of two new lizards from the Transvaal. *Ann. Mag. Nat. Hist. (ser. 6) 17*, 21-22.

Boulenger, G. A. (1896c). A list of the Reptiles and Batrachians collected by the late Prince Eugenie Ruspoli in Somaliland and Gallaland in 1893. *Ann. Mus. Civ. Stor. Nat. Genova (ser. 2) 17*, 5-16.

Boulenger, G. A. (1896d). Description of a new chamaeleon from Uganda. *Ann. Mag. Nat. Hist. (ser. 6) 17*, 376.

Boulenger, G. A. (1896e). Descriptions of new lizards from Madagascar. *Ann. Mag. Nat. Hist. (ser. 6) 17*, 444-448.

Boulenger, G. A. (1897b). Description of two new snakes. *Ann. Mag. Nat. Hist.* (*ser. 6*) *19*, 276-281.

Boulenger, G. A. (1897c). A list of the reptiles and batrachians collected in northern Nyasaland by Mr. Alex Whyte, F. Z. S., with descriptions of new species. *Proc. zool. soc. London* 800-803

Boulenger, G. A. (1897d). A list of Reptiles and Batrachians from the Congo Free State with descriptions of two new snakes. *Ann. Mag. Nat. Hist.* (*ser. 6*) *19*, 276-281.

Boulenger, G. A. (1898a). Concluding report on the late Capt. Bottego's collection of Reptiles and Batrachians from Somaliland and British East Africa. *Ann. Mus. Civ. Stor. Nat. Genova* (*ser. 2*) *18*, 715-723.

Boulenger, G. A. (1898e). On a second collection of reptiles made by Mr. E. Lort-Phillips in Somaliland. *Ann. Mag. Nat. Hist.* (*ser. 7*) *2*, 130-133.

Boulenger, G. A. (1899). Description of two new lizards from the interior of British East Africa. *Proc. zool. soc. London* 96-98.

Boulenger, G. A. (1901a). Materiaux pour la faune du Congo. Batraciens et reptiles nouveaux. *Ann. Mus. Congo, Zool* (*sect. C. ser. 1*) *2*, 7-14.

Boulenger, G. A. (1901b). A list of the batrachians and reptiles obtained by Dr. Donaldson Smith in Somaliland. *Proc. zool. soc. London* 47-49.

Boulenger, G. A. (1901c). Description of a new gecko from the Niger Delta. *Ann. Mag. Nat. Hist.* (*ser. 7*) *7*, 204.

Boulenger, G. A. (1901d). Descriptions of two new chamaeleons from Mount Ruwenzori, British East Africa. *Proc. zool. soc. London* 135-136.

Boulenger, G. A. (1902a). Description of a new snake of the genus *Psammophis* from Cape Colony. *Proc. zool. soc. London* 126.

Boulenger, G. A. (1902b). A new name for the common Agama of the Transvaal. *Ann. Mag. Nat. Hist.* (*ser. 7*) *9*, 339.

Boulenger, G. A. (1903a). Descriptions of two new lizards discovered by Mr. E. Degen on his journey to Abyssinia. *Ann. Mag. Nat. Hist.* (*ser. 7*) *11*, 54-55.

Boulenger, G. A. (1903b). On a collection of Batrachians and Reptiles from the interior of Cape Colony. *Ann. Mag. Nat. Hist.* (*ser. 7*) *12*, 215-217.

Boulenger, G. A. (1903c). Descriptions of new snakes in the collection of the British Museum. *Ann. Mag. Nat. Hist.* (*ser. 7*) *12*, 350-354.

Boulenger, G. A. (1903d). Descriptions of new lizards in the collection of the British Museum. *Ann. Mag. Nat. Hist.* (*ser. 7*) *12*, 429-435.

Boulenger, G. A. (1904b). Descriptions of three new snakes. *Ann. Mag. Nat. Hist.* (*ser. 7*) *13*, 450-452.

Boulenger, G. A. (1904c). Descriptions of two new elapine snakes from the Congo. *Ann. Mag. Nat. Hist.* (*ser. 7*) *14*, 14-15.

Boulenger, G. A. (1905c). A list of the batrachians and reptiles collected by Dr. W. J. Ansorge in Angola, with descriptions of new species. *Ann. Mag. Nat. Hist.* (*ser. 7*) *16*, 105-115.

Boulenger, G. A. (1906b). Description of a new chamaeleon of the genus *Rhampholeon* from Mashonaland. *Ann. Mag. Nat. Hist.* (*ser. 7*) *18*, 346-347.

Boulenger, G. A. (1906c). Description of a new snake of the genus *Glauconia* from Somaliland. *Ann. Mag. Nat. Hist.* (*ser. 7*) *18*, 441.

Boulenger, G. A. (1906d). Report on the reptiles collected by the late L. Fea in West Africa. *Ann. Mus. Civ. Stor. Nat. Genova* (*ser. 3*) *2*, 196-216.

Boulenger, G. A. (1906e). On a new chamaeleon from Mount Ruwenzori. *Ann. Mag. Nat. Hist.* (*ser. 7*) *18*, 473.

Boulenger, G. A. (1906f). Additions to the herptetology of British East Africa. *Proc. zool. soc. London* 570-572.

Boulenger, G. A. (1907a). On a collection of Fishes, Batrachians, and Reptiles, made by Mr. S. A. Neave in Rhodesia, north of the Zambesi, with field notes by the collector. *Mem. Proc. Lit. Philos. Soc. Manchester 51*, 1-2.

Boulenger, G. A. (1907b). Descriptions of three new lizards and a new frog discovered by Dr. W. J. Ansorge in Angola. *Ann. Mag. Nat. Hist. (ser. 7) 19*, 212-214.

Boulenger, G. A. (1907c). Descriptions of three new snakes discovered by Mr. G. L. Bates in South Cameroon. *Ann. Mag. Nat. Hist. (ser. 7) 19*, 324-326.

Boulenger, G. A. (1907d). Descriptions of two new African Lizards of the genus *Latastia*. *Ann. Mag. Nat. Hist. (ser. 7) 19*, 392-394.

Boulenger, G. A. (1907e). Descriptions of new lizards in the British Museum. *Ann. Mag. Nat. Hist. (ser. 7) 19*, 486-489.

Boulenger, G. A. (1907f). Descriptions of a new toad and a new amphisbaenid from Mashonaland. *Ann. Mag. Nat. Hist. (ser. 7) 20*, 47-49.

Boulenger, G. A. (1907g). Second report on the reptiles and batrachians collected in South Africa by Mr. C. H. B. Grant, and presented to the British Museum by Mr. C. D. Rudd. *Proc. zool. soc. London* 478-487.

Boulenger, G. A. (1908a). On a collection of freshwater fishes, batrachians and reptiles from Natal and Zululand, with descriptions of new species. *Ann. Natal Mus. 1*, 219-235.

Boulenger, G. A. (1908b). Description of three new snakes from Africa. *Ann. Mag. Nat. Hist. (ser. 8) 2*, 93-94.

Boulenger, G. A. (1909a). List of Reptiles collected by Capt. G. Ferrari at Jumbo, lower Juba. *Ann. Mus. Civ. Stor. nat. Giacomo Doria 44*, 308-309.

Boulenger, G. A. (1909b). List of Reptiles and Batrachians collected by Capt. U. Ferrandi at Bardera. *Ann. Mus. Civ. Stor. nat. Giacomo Doria 44*, 310-311.

Boulenger, G. A. (1909c). A list of the freshwater Fishes, Batrachians, and Reptiles obtained by Mr. J. Stanley Gardiner's Expedition to the Indian Ocean. *Trans. Linn. Soc. London (ser. 2) 12*, 291-301.

Boulenger, G. A. (1910a) A revised list of the South African reptiles and batrachians, with synoptic tables, special reference to the specimens in the South African Museum, and descriptions of new species. *Ann. S. Afr. Mus. 5*, 455-538.

Boulenger, G. A. (1910b). Description of four new African snakes. *Ann. Mag. Nat. Hist. (ser. 8) 5*, 512-513.

Boulenger, G. A. (1911a). Descriptions of three new snakes discovered by Mr. G. L. Bates in South Cameroon. *Ann. Mag. Nat. Hist. (ser. 8) 8*, 370-371.

Boulenger, G. A. (1911c). On a third collection of reptiles and batrachians made by Dr. E. Bayon in Uganda. *Ann. Mus. Civ. Stor. nat. Genova (ser. 3) 5*, 161-189.

Boulenger, G. A. (1912b). Missione per la Frontiere Italo-Etiopica sotto il Comando del Capitano Carlo Citerni. Resultati zoologici. *Ann. Mus. Civ. Stor. nat. Genova (ser. 3) 5*, 329-332.

Boulenger, G. A. (1913a). Description d'une reptile amphisbenide nouveau provenant du Katanga. *Rev. Zool. Afr. (Brussels) 2*, 392-393.

Boulenger, G. A. (1913b). Description de deux Reptiles nouveaux provenant du Katanga. *Revue Zool. Afr. 3*, 103-105.

Boulenger, G. A. (1914). Descriptions of new species of snakes in the collection of the British Museum. *Ann. Mag. Nat. Hist. (ser. 8) 14*, 482-485.

Boulenger, G. A. (1915). A list of the snakes of the Belgian and Portuguese Congo, Northern Rhodesia and Angola. *Proc. zool. soc. London* 193-223.

Boulenger, G. A. (1916a). On the lizards allied to *Lacerta agilis* and *Lacerta parva*. *Trans. zool. soc. London 21*, 1-90.

Boulenger, G. A. (1917a). On the lizards of the genus *Philochortus* Matschie. *Proc. zool. soc. London* 145-157.

Boulenger, G. A. (1918c). Sur les lezards du genre *Acanthodactylus* Wieg. *Bull. soc. zool. France 43*, 143-155.

Boulenger, G. A. (1918d). A synopsis of the lizards of the genus *Eremias. J. Zool. Res. London 3*, 1-12.

Boulenger, G. A. (1919). Batraciens et reptiles recueillis par le Dr. C. Christy au Congo Belge dans les Districts de Stanleyville, Haut-Uele et Ituri en 1912-1913. *Revue Zool. Afr. 7*, 1-29.

Boulenger, G. A. (1920). *Monograph of the Lacertidae, I.* London, pp. 352.

Boulenger, G. A. (1921). *Monograph of the Lacertidae, II.* London,. pp. 451.

Boulenger, G. A. and Power, J. H. (1921). A revision of the South African Agamas allied to *Agama hispida* and *A. atra. Trans. Roy. Soc. S. Afr. 9*, 229-287.

Bourgat, R. (1967). Introduction a l'etude ecologique du *C. pardalis* Cuv. de l'ile de la Reunion. *Vie et Milieu 18*, 221-230.

Bourgat, R. (1970). Recherches ecologiques et biologiques sur le *Chamaeleo pardalis* Cuvier 1829 de l'ile de la Reunion et de Madagascar. *Bull. soc. zool. France 95*, 259-268.

Bourgat, R. and Brygoo, E. R. (1968). Apport de l'etude des hemipenis a la systematique du complexe *C. verrucosus* Cuv., *C. oustaleti* Mocquard. *Ann. Fac. Sci. Univ. Madagascar 6*, 418-424.

Bourgeois, M. (1963). Note sur *Atractaspis irregularis conradsi* Sternfeld (Viperidae). Structure de crane et de l'apparail de la morsure. *Ann. Soc. r. zool. Belg. 93*, 159-169.

Bourgeois, M. (1968). Contribution a la morphologie comparee du crane des ophidiens de l'Afrique Centrale. *Publ. Univ. Off. Congo Lubumbashi 18*, 1-293.

Bourquin, O. (1977a). Some reptile records from northeastern Zululand. *Lammergeyer 23*. 46-47.

Bourquin, O. (1977b). The Transvaal montane skink—a new record for Natal. *Lammergeyer 23*, 48.

Boycott, Richard C. and Haacke, W. D. (1979). Note on the type-locality, distribution and juvenile coloration of *Naja nigricollis woodi* (Serpentes: Elapidae) and on account of the colour-pattern variation in intergrade populations. *Ann. Cape Prov. Mus. Nat. Hist. 13*, 31-38.

Bradley, O. C. (1903). The muscles of mastication and the movements of the skull in Lacertilia. *Zool. Jb. Abt. Anat. 18*, 475-488.

Brain, C. K. (1958). Web-footed geckos of the Namib (*Palmatogecko rangei* Andersson). *Afr. Wild Life 12*, 67-70.

Brain, C. K. (1959a). Notes on the breeding of the South African skink, *Typhlosaurus lineatus* Boulenger. *Copeia* 70-71.

Brain, C. K. (1959b). Mating in the South African Mole snake *Pseudaspis cana* (Linnaeus). *Copeia* 71-72.

Brain, C. K. (1960). On the hemipenis of the Mole snake *Pseudaspis cana* (Linn.) *Copeia* 155.

Brain, C. K. (1961). *Chamaeleo dilepis*—a study of its biology and behavior. *J. Herpet. Ass. Rhodesia 15*, 15-20.

Brain, C. K. (1962a). Observations on the birth of young *Chamaeloe jacksoni. J. Herpet. Ass. Rhodesia 17-18*, 11-12.

Brain, C. K. (1962b). A review of the gecko genus *Ptenopus* with the description of a new species. *Cimbebasia 1*, 1-18.

Brain, C. K. (1962c). Observations on the temperature tolerance of lizards in the Central Namib Desert, South West Africa. *Cimbebasia 4*, 1-5.

Branch, W. R. (1973). Birth in the mole snake (*Pseudaspis cana*). *J. Herpet. Ass. Africa 10*, 24.

Branch, W. R. (1976). The Wolf snakes *Lycophidion capense* and *Lycophidion variegatum* (Reptilia, Serpentes, Colubridae) in South Africa. *J. Herpet. 10,* 1-11.

Branch, W. R. (1979). Dicephalism in the African Worm snake *(Leptotyphlops bicolor.* Herpet. Review 10, 6-7.

Branch, W. R. and Erasmus, H. (1976). Reproduction in Madagascar ground and tree boas, *Acrantophis madagascariensis* and *Sanzinia madagascariensis. Inter. Zoo Yrbk.16,* 78-80.

Branch, W. R. and Patterson, R. W. (1975). Notes on the development of embryos of the African Rock Python *Python sebae* (Serpentes, Boidae). *J. Herpet. 9,* 243-248.

Briscoe, M. S. (1949). Notes on snakes collected in Liberia. *Copeia* 16-18.

Britt, David P. (1978). Death following the bite of a burrowing viper. *Nigerian Field 43,* 41-42.

Broadley, D. G. (1956). Snakes of Southern Rhodesia. 1: The Striped Skaapsteker or Three-lined snake, *Psammophylax* (formerly *Trimerorhinus*) *tritaeniatus. Afr. Wild Life 10,* 215-216.

Broadley, D. G. (1957a). Snakes of Southern Rhodesia. 2: The Northern Green Snake, *Philothamnus irregularis irregularis* (Leach). *Afr. Wild Life 11,* 53-55.

Broadley, D. G. (1957b). Snakes of Southern Rhodesia. 3: The Night Adders, genus *Causus. Afr. Wild Life 11,* 115-118.

Broadley, D. G. (1957c). Snakes of Southern Rhodesia. 4: The Southeastern Vine or Twig Snake, *Thelotornis kirtlandii capensis. Afr. Wild Life 11,* 297-300.

Broadley, D. G. (1958a). Snakes of Southern Rhodesia. 5: The Egg-eater. *Dasypeltis scabra. Afr. Wild Life 12,* 29-32.

Broadley, D. G. (1958b). Snakes of Southern Rhodesia. 6: The Egyptian cobra *(Naja haje haje). Afr. Wild Life 12,* 125-128.

Broadley, D. G. (1958c). Snakes of Southern Rhodesia. 7: The Common House snake *(Boaedon fuliginosus fuliginosus). Afr. Wild Life 12,* 197-199.

Broadley, D. G. (1958d). Serpentes, Colubridae: *Duberria lutrix rhodesiana* n. subsp. *Occ. Pap. natn. Mus. S. Rhod.* (ser. B) 3, 215-216.

Broadley, D. G. (1959a). Snakes of Southern Rhodesia. 8: The Olive Grass snake *Psammophis sibilans sibilans* (Linnaeus). *Afr. Wild Life 13,* 29-31.

Broadley, D. G. (1959b). The herpetology of Southern Rhodesia. I. Snakes. *Bull. Mus. comp. Zool. Harvard 120,* 1-100.

Broadley, D. G. (1959c). Snakes of Southern Rhodesia. 9: The Purple-glossed Snake *(Calamelaps unicolor miolepis* Gunther). *Afr. Wild Life 13,* 201-203.

Broadley, D. G. (1959d). A review of the Green-snakes, genus *Philothamnus,* of the Rhodesias. *Occ. Pap. natn. Mus. S. Rhod.* (ser. B) 3, 309-313.

Broadley, D. G. (1959e). Sauria, Cordylidae: *Platysaurus guttatus pungweensis* n. subsp. *Occ. Pap. natn. Mus. S. Rhod.* (ser. B) 3, 314-315.

Broadley, D. G. (1960a). Serpentes, Colubridae: *Psammophylax tritaeniatus fitzgeraldi* n. subsp. *Occ. Pap. natn. Mus. S. Rhod.* (ser. B) 3, 431-432.

Broadley, D. G. (1960b). Sauria, Cordylidae: *Gerrhosaurus nigrolineatus anselli* n. subsp. *Occ. Pap. natn. Mus. S. Rhod.* (ser. B) 3, 433-434.

Broadley, D. G. (1960c). Sauria, Scincidae: *Mabuya damarana rhodesiana* n. subsp. *Occ. Pap. natn. Mus. S. Rhod.* (ser. B.) 3, 435-436.

Broadley, D. G. (1961a). Snakes of Southern Rhodesia. 10: The File Snakes *(Mehelya). Afr. Wild Life 15,* 72-74.

Broadley, D. . (1961b). The African Python. *Black Lechwe 3,* 32-34.

Broadley, D. G. (1961c). Snakes of Southern Rhodesia: 11: The Black Mamba *(Dendroaspis polylepis polylepis). Afr. Wild Life 15,* 299-302.

Broadley, D. G. (1962a). Snakes of Southern Rhodesia: 12: The Common Boomslang *(Dispholidus typus typus). Afr. Wild Life 16,* 109-111.

Broadley, D. G. (1962b). Serpentes, Colubridae: *Natriciteres olivaceus bipostocularis* n. subsp. *Occ. Pap. natn. Mus. S. Rhod. (ser. B) 3*, 785-786.

Broadley, D. G. (1962c). On some reptile collections from the northwestern and northeastern district of Southern Rhodesia 1958-1961, with descriptions of four new lizards. *Occ. Pap. natn. Mus. S. Rhod. (ser. B) 3*, 787-843.

Broadley, D. G. (1962d). Notes on *Chamaeleo marshalli. J. Herpet. Ass. Rhodesia 17-18*, 18.

Broadley, D. G. (1963a). Two rare fossorial reptiles in southeastern Rhodesia — predator and prey. *J. Herpet. Ass. Rhodesia 20*, 7-8.

Broadley, D. G. (1963b). An expedition to the Mozambique plain, November 1962. *J. Herpet. Ass. Rhodesia 20*, 14-18.

Broadley, D. G. (1963c). Snakes of Southern Rhodesia. 13: The Stripe-bellied Sand snake (*Psammophis subtaeniatus subtaeniatus*). *Afr. Wild Life 17*, 39-40.

Broadley, D. G. (1963d). Three new lizards from south Nyasaland and Tete. *Ann. Mag. Nat. Hist. (ser. 13) 6*, 285-288.

Broadley, D. G. (1964a). A report on the opportunity for zoological collecting afforded by the Beira-Feruka pipe-line project. *Herpet. Ass. Rhod. Newsletter 1*, 1-3.

Broadley, D. G. (1964b). A new species of *Platysaurus* from central Mozambique. *Arnoldia Rhodesia 1*(5), 1-2.

Broadley, D. G. (1964c). A review of the crag-lizards (genus *Pseudocordylus*) of Natal. *Ann. Natal Mus. 16*, 99-110.

Broadley, D. G. (1965a). A new chamaeleon from Malawi. *Arnoldia Rhodesia 1*(32), 1-3.

Broadley, D. G. (1965b). A new species of *Platysaurus* from northern Mozambique. *Arnoldia Rhodesia 1*(33), 1-4.

Broadley, D. G. (1965c). A revision of the *Prosymna sundevalli* group (Serpentes, Colubridae). *Arnoldia Rhodesia 2*(5), 1-6.

Broadley, D. G. (1966a). A review of the *Riopa sundevalli* group (Sauria: Scincidae) in southern Africa. *Arnoldia Rhodesia 2*(34), 1-7.

Broadley, D. G. (1966b). A review of the genus *Natriciteres* Loveridge (Serpentes: Colubridae). *Arnoldia Rhodesia 2*(35), 1-11.

Broadley, D. G. (1966c). A review of the African Stripe-bellied Sand-snakes of the genus *Psammophis. Arnoldia Rhodesia 2*(36), 1-9.

Broadley, D. G. (1966d). A review of the Natal Green snake *Philothamnus natalensis* (A. Smith), with a description of a new subspecies. *Ann. Natal Mus. 18*, 417-423.

Broadley, D. G. (1967a). A review of the genus *Lycodonomorphus* Fitzinger (Serpentes: Colubridae) in southeastern Africa, with a key to the genus: *Arnoldia Rhodesia 3*(16), 1-9.

Broadley, D. G. (1967b). A new species of *Ichnotropis* (Sauria: Lacertidae) from the Botswana-Caprivi border. *Arnoldia Rhodesia 3*(24), 1-5.

Broadley, D. G. (1967c). The life cycles of two sympatric species of *Ichnotropis* (Sauria: Lacertidae). *Zool. Africana 3*, 1-2.

Broadley, D. G. (1968a). A review of the African cobras of the genus *Naja* (Serpentes: Elapinae). *Arnoldia Rhodesia 3*(29), 1-14.

Broadley, D. G. (1968b). A revision of the African genus *Typhlosaurus* Wiegmann (Sauria: Scincidae). *Arnoldia Rhodesia 3*(36), 1-20.

Broadley, D. G. (1968c). A revision of *Aspidelaps scutatus* (A. Smith) (Serpentes: Elapinae). *Arnoldia Rhodesia 4*(2), 1-9.

Broadley, D. G. (1968d). The African Wolf Snakes (Serpentes: Colubridae): A new genus and species from Mozambique. *Arnoldia Rhodesia 4*(10), 1-5.

Broadley, D. G. (1968e). A new species of *Crotaphopeltis* (Serpentes: Colubridae) from Barotseland, Zambia. *Fieldiana Zool. 51*, 135-139.

Broadley, D. G. (1969). A new species of *Lycophidion* from Rhodesia (Serpentes; Colubridae). *Arnoldia Rhodesia* 4(27), 1-8.

Broadley, D. G. (1971a). A review of *Rhamphiophis acutus* (Gunther) with the description of a new subspecies from Zambia (Serpentes: Colubridae). *Arnoldia Rhodesia* 5(8), 1-8.

Broadley, D. G. (1971b). A review of *Rhampholeon marshalli* Boulenger, with the description of a new subspecies from Mozambique (Sauria: Chamaeleonidae). *Arnoldia Rhodesia* 5(10), 1-6.

Broadley, D. G. (1971c). A revision of the African snake genus *Elapsoidea* Bocage (Elapidae). *Occ. Pap. Natl. Mus. Rhodesia (ser. B)* 4, 577-626.

Broadley, D. G. (1971d). A revision of the African snake genera *Amblyodipsas* and *Xenocalamus* (Colubridae). *Occ. Pap. Natl. Mus. Rhodesia (ser. B)* 4, 629-697.

Broadley, D. G. (1971e). The reptiles and amphibians of Zambia. *Puku* 6, 1-143.

Broadley, D. G. (1972a). A review of the *Nucras tessellata* group (Sauria: Lacertidae). *Arnoldia Rhodesia* 5(20), 1-36.

Broadley, D. G. (1972b). The Horned Viper, *Bitis caudalis* (A. Smith) in the central Kalahari. *Botswana Notes Rec.* 4, 263-264.

Broadley, D. G. (1973). Addenda and Corrigenda to 'The reptiles and amphibians of Zambia'. *Puku* 7, 93-95.

Broadley, D. G. (1974a). A review of the *Mabuya maculilabris* group in southeastern Africa (Sauria: Scincidae). *Arnoldia Rhodesia* 6(23), 1-10.

Broadley, D. G. (1974b). A puzzling case of ophiophagy in Rhodesia. *J. Herpet.* 8, 247.

Broadley, D. G. (1974c). A review of the cobras of the *Naja nigricollis* complex in southwestern Africa (Serpentes: Elapidae). *Cimbebasia (ser. A)* 2, 155-162.

Broadley, D. G. (1974d). Rhodesian Reptiles no. 3. (*Atractaspis bibroni; Lycophidion variegatum*). *Wild Rhodesia* 4, 21.

Broadley, D. G. (1974e). Reproduction in the genus *Platysaurus* (Sauria: Cordylidae). *Herpetologica* 30, 379-380.

Broadley, D. G. (1975a). A review of *Psammophis leightoni* and *Psammophis notostictus* in southern Africa (Serpentes: Colubridae). *Arnoldia Rhodesia* 7(13), 1-17.

Broadley, D. G. (1975b). A review of the *Mabuya lacertiformis* complex in Southern Africa (Sauria: Scincidae). *Arnoldia Rhodesia* 7(18), 1-16.

Broadley, D. G. (1976). Two new forms of *Platysaurus* from the northern Transvaal. *Arnoldia Rhodesia* 8(8), 1-3.

Broadley, D. G. (1977a). A revision of the African snakes of the genus *Psammophylax* Fitzinger (Colubridae). *Occ. Pap. Natl. Mus. Rhodesia (ser. B)* 6, 1-44.

Broadley, D. G. (1977b). A review of the genus *Psammophis* in southern Africa (Serpentes: Colubridae). *Arnoldia Rhodesia* 8(12), 1-29.

Broadley, D. G. (1977c). A review of the northeastern forms of the *Pachydactylus capensis* complex (Sauria: Gekkonidae). *Arnoldia Rhodesia* 8(18), 1-19.

Broadley, D. G. (1977d). A review of the *Hemidactylus mabouia* complex in southeastern Africa (Sauria: Gekkonidae). *Arnoldia Rhodesia* 8(19), 1-15.

Broadley, D. G. (1978). A revision of the genus *Platysaurus* A. Smith (Sauria: Cordylidae). *Occ. Pap. Natl. Mus. Rhodesia (ser. B)* 16, 129-185.

Broadley, D. G. (1979). Predation on reptile eggs by African snakes of the genus *Prosymna. Herpetologica* 35, 338-341.

Broadley, D. G. and Cock, E. V. (1975). *Snakes of Rhodesia.* Longman Rhodesia, Salisbury.

Broadley, D. G. and Gans, C. (1969). A new species of *Zygaspis* (Amphisbaenia: Reptilia) from Zambia and Angola. *Arnoldia Rhodesia* 4(25), 1-4.

Broadley, D. G. and Gans, C. (1978a). Southern forms of *Chirindia* (Amphisbaenia, Reptilia). *Ann. Carnegie Mus.* 47, 29-51.

Broadley, D. G. and Gans, C. (1978b). Distribution, variation and systematic status of *Zygaspis violacea* (Peters) (Amphisbaenia: Reptilia) endemic to southeastern Africa. *Ann. Carnegie Mus.* 47, 319-334.

Broadley, D. G., Gans, C. and Visser, J. (1976). Studies on Amphisbaenians (Amphisbaenia, Reptilia). 6. The genera *Monopeltis* and *Dalophia* in southern Africa. *Bull. Amer. Mus. Nat. Hist.* 157, 311-486.

Broadley, D. G. and Greer, A. E. (1969). A revision of the genus *Acontais* Cuvier (Sauria: Scincidae). *Arnoldia Rhodesia* 4(26), 1-29.

Broadley, D. G. and Parker, Raymond H. (1976). Natural hybridization between the puff adder and gaboon viper in Zululand (Serpentes: Viperidae). *Durban Mus. Novitates* 11, 77-83.

Broadley, D. G. and Stevens, R. A. (1971). A review of *Chamaetortus aulicus* Gunther, with the description of a new subspecies from Malawi (Serpentes: Colubridae). *Arnoldia Rhodesia* 5(11), 1-11.

Broadley, D. G. and Watson, Gillian. (1976). A revision of the worm snakes of southeastern Africa (Serpentes: Leptotyphlopidae). *Occ. Pap. Natl. Mus. Rhodesia (ser. B)* 5, 465-510.

Brock, G. T. (1929). The development of the skull of *Leptodeira hotamboia*. *Quart. J. micro. Sci.* 73, 289-334.

Brock, G. T. (1932a). The skull of *Leptotyphlops (Glauconia) nigricans*. *Anat. Anz.* 73, 199-204.

Brock, G. T. (1932b). Some developmental stages in the skull of the geckos, *Lygodactylus capensis* and *Pachydactylus maculosa*, and their bearing on certain important problems in Lacertilian Craniology. *S. Afr. J. Sci.* 29, 508-532.

Brock, G. T. (1940). The skull of the chamaeleon, *Lophosaura ventralis* Gray; some developmental stages. *Proc. zool. soc. London* 9, 219-241.

Brock, G. T. (1941). The skull of *Acontias meleagris*, with a study of the affinities between lizards and snakes. *Zool. J. Linn. Soc.* 41, 71-88.

Brongersma, L. D. (1967). Guide for the identification of stranded turtles on British Coasts, *Brit. Mus. (Nat. Hist.)* 659, 1-22.

Bruno, S. and Hotz, H. (1976). *Coluber hippocrepis* auf der Insel Sardinien (Reptilia, Serpentes, Colubridae). *Salamandra* 12, 69-86.

Bruton, M. N. (1977). Feeding, social behaviour and temperature preferences in *Agama atra* Daudin (Reptilia, Agamidae). *Zool. Africaine* 12, 183-199.

Brygoo, E. R., Blanc, C. P. and Domergue, C. A. (1970a). Notes sur les *Brookesia* de Madagascar: III. *Brookesia karchei* n. sp. du Massif du Marojezy. *Ann. Fac. Sci. Univ. Madagascar* 7, 267-271.

Brygoo, E. R., Blanc, C. P. and Domergue, C. A. (1970b). Notes sur les *Chamaeleo* de Madagascar: IV. *C. gastrotaenia* n. subsp. du Massif du Marojezy. *Ann. Fac. Sci. Univ. Madagascar* 7, 273-278.

Brygoo, E. R. and Domergue, C. A. (1966). Notes sur *Chamaeleo willsi* Gunther 1890 et description d'une sous-espece nouvelle: *C. willsi petteri* n. ssp. *Bull. Mus. nat. Hist. nat. Paris (ser. 2)* 38, 353-361.

Brygoo, E. R. and Domergue, C. A. (1967). Description d'un Cameleon nouveau de Madagascar *Chamaeleo tsaratananensis* n. sp. *Bull. Mus. nat. Hist. nat. Paris (ser. 2)* 39, 829-832.

Brygoo, E. R. and Domergue, C. A. (1968a). Notes sur les *Brookesia* de Madagascar: I. Description d'un nouveau *Brookesia* de Madagascar: *B. vadoni* n. sp. (Chamaeleontide). *Bull. Mus. nat. Hist. nat. Paris (ser. 2)* 40, 677-682.

Brygoo, E. R. and Domergue, C. A. (1968b). Description du male de *C. willsi petteri* Brygoo et Domergue, Cameleon du nord de Madagascar. *Bull. Mus. nat. Hist. nat. Paris (ser. 2) 40,* 891-896.

Brygoo, E. R. and Domergue, C. A. (1968c). Notes sur les *Brookesia* de Madagascar: II. Un *Brookesia* des forets orientales de Madagascar, *B. thieli* n. sp. (Chamaeleonides). *Bull. Mus. nat. Hist. nat. Paris* (ser. 2) *40,* 1103-1109.

Brygoo, E. R. and Domergue, C. A. (1969). Notes sur les *Chamaeleo* de Madagascar: IV. *Chamaeleo guentheri* Boulenger 1888 synonyme de *C. pardalis* Cuvier 1829. *Bull. Mus. nat. Hist. nat. Paris (ser. 2) 41,* 117-121.

Burchfield, P. M. (1975). Hatching the radiated tortoise, *Testudo radiata,* at Brownsville Zoo. *Inter. Zoo Yrbk. 15,* 90-92.

Burger, W. L. and Natsuno, T. (1974). A new genus for the Arafura Smooth seasnake and redefinitions of other seasnake genera. *The Snake 6,* 61-75.

Burns, B. and Pickwell, G. V. (1972). Cephalic glands in sea snakes (*Pelamis, Hydrophis* and *Laticauda*). *Copeia* 547-558.

Burrage, B. R. (1973). Comparative ecology and behaviour of *Chamaeleo pumilis pumilis* (Gmelin) and *C. namaquensis* A. Smith (Sauria: Chamaeleonidae). *Ann. S. Afr. Mus. 61,* 1-158.

Burrage, B. R. (1974). Population structure in *Agama atra* and *Cordylus cordylus cordylus* in the vicinity of De Kelders, C. P. *Ann. S. Afr. Mus. 66,* 1-23.

Busack, Stephen D. (1975). Biomass estimates and thermal environment of a population of the fringe-toed lizard *Acanthodactylus pardalis. Brit. J. Herpet. 5,* 457-459.

Bustard, H. Robert. (1958). Use of horns by *Chamaeleo jacksoni. Brit. J. Herpet. 2,* 105-107.

Bustard, H. Robert (1963a). The three-horned chamaeleon (*C. jacksoni*). *Aquar. and Pondkpr. 28,* 6-7.

Bustard, H. Robert (1963b). Notes on the eyed gecko (*Pachydactylus geitje*) with special reference to incubation. *Copeia* 433-434.

Bustard, H. Robert (1965). Observations on the life history and behavior of *Chamaeleo hoehnelii* Steindachner. *Copeia* 401-410.

Bustard, H. Robert (1966). Observations on the life history and behavior of *Chamaeleo bitaeniatus* Fischer. *Herpetologica 22,* 13-23.

Bustard, H. Robert (1967). Comparative behavior of chamaeleons: fight behavior in *Chamaeleo gracilis* Hallowell. *Herpetologica 23,* 44-50.

Bustard, H. Robert (1971). Temperature and water tolerances of incubating Sea Turtle eggs. *Brit. J. Herpet. 4,* 196-198.

Bustard, H. Robert (1972). *Sea Turtles, Natural History and Conservation.* Collins, London, pp. 220.

Bustard, H. Robert and Greenham, Peter (1968). Physical and chemical factors affecting hatching in the green sea turtle, *Chelonia mydas* (L.). *Ecology 49,* 269-276.

Calabresi, Enrica (1915). Contributo alla Conoscenza dei Rettili della Somalia. *Monitore zool. ital. 26,* 234-247.

Calabresi, Enrica (1918). Rettili somali raccolti fra Gelib e Margherita (Riva sinistra del Guiba) dai Dott. Mazzocchi e Scarsellati, nell 1912. *Monitore zool. ital. 29,* 122-124.

Calabresi, Enrica (1923a). Missione Zoologica del Dr. E. Festa in Cirenaica. *Boll. Mus. Zool. Anat. Comp. Univ. Torino 38*(7), 1-28.

Calabresi, Enrica (1923b). Anfibi e Rettili dell'Afria orientale raccolti durante le Spedizioni Franchetti e Zammarano. *Atti. Soc. Ital. Sci. Nat. Milan 62,* 145-163.

Calabresi, Enrica (1925). Anfibi e Rettili raccolti dal Signor Ugo Ignesti nell' Abissinia settentrionale. *Atti. Soc. Ital. Sci. Nat. Milan 64,* 100-109.

Calabresi, Enrica (1927). Anfibi e rettili raccolti nella Somalia dai Proff. G. Stefanini e N. Puccioni (Gennaio—Luglio 1924). *Atti. Soc. Ital. Sci. Nat. Milan 66,* 14-60.

Caldwell, David K., Carr, Archie and Ogren, Larry (1959). The Atlantic loggerhead sea turtle, *Caretta caretta caretta* (L.) in America. I. Nesting and migration of the Atlantic loggerhead turtle. *Bull. Florida State Museum 4,* 295-308.

Camp, Charles L. (1923). Classification of the lizards. *Bull. Amer. Mus. Nat. Hist. 48,* 289-481. (Facsimile reprint: 1970., S. S. A. R.).

Campbell, Jonathan A. and Murphy, James B. (1977). Miscellaneous notes on the reproductive biology of reptiles. I. Two colubrid snake species from the Malagasy Republic, *Leioheterodon madagascariensis* and *Madagascariophis colubrina* (Reptilia, Serpentes, Colubridae). J. Herpet. 11, 228-230.

Cansdale, G. S. (1961). *West African Snakes.* Longmans, London pp. 74.

Carlsson, F. H. H. (1974). Snake venom toxins. The primary structures of two novel cytotoxin homologues from the venom of forest cobra (*Naja melanoleuca*). *Biochem. biophys. Res. Commun. 59,* 269-276.

Carlsson, F. H. H. and Louw, A. I. (1978). The oxidation of methionine and its effect on the properties of cardiotoxin $V^{II}1$ from *Naja melanoleuca* venom. *Biochim. Biophys. Acta 534,* 322-330.

Carpenter, C. C., Murphy, J. B. and Mitchell, L. A. (1978). Combat bouts with spur use in the Madagascan boa (*Sanzinia madagascariensis*). *Herpetologica 34,* 207-212.

Carpenter, G. D. H. (1925). *A Naturalist in East Africa.* Oxford pp. 187.

Carr, A. and Ogren, L. (1959). The ecology and migrations of sea turtles. 3. *Dermochelys* in Costa Rica. *Amer. Mus. Novitates 1958,* 1-29.

Cerf, F. le (1907). Reptiles et Batraciens observes a Maison-Caree (Algerie). *Ann. Assoc. Natur. Levallois-Perret 13,* 22-26.

Chabanaud, P. (1916g). Enumeration des Ophidiens non encore etudies de l'Afrique occidentale, appartenant aux collections du Museum avec le description des especes et des varietes nouvelles. *Bull. Mus. nat. Hist. nat. Paris 22,* 362-382.

Chabanaud, P. (1917a). Description d'un Lacertilien nouveau de Maroc. *Bull. Mus. nat. Hist. nat. Paris 23,* 3-6.

Chabanaud, P. (1917b). Note complementaire sur les Ophidiens de l'Afrique occidentale, appartenant aux Collections du Museum, avec la description d'une espece nouvelle. *Bull. Mus. nat. Hist. nat. Paris 23,* 7-14.

Chabanaud, P. (1917c). Enumeration des reptiles non encore etudies de l'Afrique occidentale, appartenant aux Collections du Museum, avec le description des especes nouvelles. *Bull. Mus. nat. Hist. nat. Paris 23,* 83-105.

Chabanaud, P. (1917e). Descriptions de trois especes nouvelles de Reptiles de l'Afrique. *Bull. Mus. nat. Hist. nat. Paris 23,* 219-225.

Chabanaud, P. (1917f). Etude complementaire sur les Lacertilians de l'Afrique occidentale. *Bull. Mus. nat. Hist. nat. Paris 23,* 226-228.

Chabanaud, P. (1917h). Revision de quelques Reptiles d'Afrique et description de trois especes nouvelles. *Bull. Mus. nat. Hist. nat. Paris 23,* 442-454.

Chabanaud, P. (1918a). Etude complementaire de deux *Agama* de l'Afrique occidentale et description de quatre especes nouvelles de reptiles de la meme region. *Bull. Mus. nat. Hist. nat. Paris 24,* 104-112.

Chabanaud, P. (1918b). Etude d'une collection de reptiles de l'Afrique Occidentale francaise, recemment donne au Museum d'Histoire naturelle de Paris par le Dr. G. Bouet, avec la description de deux especes nouvelles. *Bull. Mus. nat. Hist. nat. Paris 24,* 160-166.

Chabanaud, P. (1921). Contribution a l'etude de la faune herpetologique de l'Afrique Occidentale. *Bull. Com. Etudes Hist. Sci. Afr. Occ. franc.,* 445-472.

Chapman, B. M. and Chapman R. F. (1964). Observations on the biology of the lizard *Agama agama* in Ghana. *Proc. zool. soc. London 143,* 121-132.

Chapman, D. S. (1968). The symptomatology, pathology and treatment of the bites of venomous snakes of central and southern Africa. in, *Venomous Animals and their venoms* (W. Bucherl, E. E. Buckley and V. Deulofeu, editors), Academic Press, London and New York. *1,* 468-527.

Charnier, M. (1965). Le cycle sexuel chez le lezard male *Agama agama* (Reptilia, Lacertilien) dans la region de Dakar. *Ann. Fac. Sci. Darkar 18,* 33-59.

Cheymol, J., Boquet, P., Bourillet, F., Detrait, J. and Roch-Arveiller, M. (1973). Comparaison des principales proprietes pharmacologiques de differents venins d'*Echis carinatus* (viperides). *Archs. int. Pharmacodyn. Ther. 205,* 293-304.

Chippaux, J. P., N'Guessan, G., Paris, F. X., Roland, G. and Kebe, M. (1978). Spitting cobra (*Naja nigricollis*) bite. *Trans. R. Soc. Trop. med. Hyg. 72,* 106.

Cloudsley-Thompson, J. L. (1965). Rhythmic activity, temperature-tolerance, water-relations and mechanism of heat death in a tropical skink and gecko. *J. Zool. London, 146,* 55-69.

Cloudsley-Thompson, J. L. (1967). Water-relations and diurnal rhythm of activity in the young Nile monitor. *Brit. J. Herpet. 3,* 296-300.

Cloudsley-Thompson, J. L. (1969). Diurnal rhythm of activity in the Nile monitor. *Brit. J. Herpet. 4,* 112-114.

Cloudsley-Thompson, J. L. (1970). On the biology of the desert tortoise *Testudo sulcata* in Sudan. *J. Zool. London 160,* 17-33.

Cloudsley-Thompson, J. L. (1972). Site tenure and selection in the African gecko *Tarentola annularis* (Geoffroy). *Brit. J. Herpet. 4,* 286-292.

Cloudsley-Thompson, J. L. and Chadwick, M. J. (1964). *Life in deserts.* Foulis, London pp. 218.

Cogger, H. G. (1966). The status of the 'elapid' snake *Tropidechis dunensis* De Vis. *Copeia* 893.

Cole, L. R. (1967). The snake *Miodon acanthias* found with *Geotrypetes seraphini* (Amphibia: Caeciliidae) as prey. *Copeia* 862.

Conant, R. (1937). Note on eggs and youngs of *Lioheterodon madagascariensis* (D. and B.) *Zoologica, N.Y. 4,* 23.

Cooper, J. E. (1974). Parasites from reptiles in Kenya with notes on their significance and control. *Brit. J. Herpet. 5,* 431-438.

Cope, E. D. (1861). Catalogue of the Colubridae in the Museum of the Academy of Natural Sciences of Philadelphia. Part 3. *Proc. Acad. Philadelphia 1860,* 553-566.

Cope, E. D. (1862a). On *Lacerta echinata* and *Tiliqua dura. Proc. Acad. nat. Sci. Philadelphia 7,* 189-191.

Cope, E. D. (1862b). Notes upon some reptiles of the Old World. *Proc. Acad. nat. Sci. Philadelphia 7,* 337-344.

Cope, E. D. (1868). Observations on reptiles of the Old World. Art. II. *Proc. Acad. nat. Sci. Philadelphia* 314-323.

Cope, E. D. (1869). Seventh contribution to the herpetology of tropical America. *Proc. Amer. Phil. Soc. 11,* 147-169.

Cope, E. D. (1885). Twelfth contribution to the herpetology of tropical America. *Proc. Amer. Phil. Soc. 22,* 167-194.

Corkill, N. L. and Kirk, R. (1954). Poisoning by the Sudan mole viper *Atractaspis microlepidota* Gunther. *Trans. R. Soc. Trop. Med. Hyg. 48,* 376-384.

Cott, H. (1934). The Zoological Society's expedition to the Zambesi, 1927. V. On a collection of lizards, mainly from East Africa, with descriptions of new species of *Zonurus, Monopeltis* and *Chirindia. Proc. zool. soc. London 5,* 145-173.

Cott, H. B. (1961). Scientific results of an inquiry into the ecology and economic status of the Nile crocodile (*Crocodylus niloticus*) in Uganda and Northern Rhodesia. *Trans. zool. soc. London 29*, 211-356.

Cowles, R. B. (1930). The life history of *Varanus niloticus* Linnaeus as observed in Natal, South Africa. *J. Entomol. Zool. 22*, 1-31.

Cowles, R. B. (1956). Notes on natural history of South African agamid lizards. *Herpetologica 12*, 297-302.

Crawford, Caroline M. and Thorpe, Roger S. (1979). Body temperatures of two geckos (*Phelsuma*) and a skink (*Mabuya*) in Praslin, Seychelles. *Brit. J. Herpet. 6*, 25-31.

Crisp, Mary; Cook, L. M. and Hereward, F. V. (1979). Color and heat balance in the lizard *Lacerta dugesii*. *Copeia* 250-258.

Csiki, E. (1903). 'no title'. *Rovartani Lapok 10*, 198.

Csiki, E. (1904). Ueber einige Gattungsnamen. *Zool. Anz. 28*, 266-267.

Curry-Lindahl, K. (1957). Behaviour of the tropical rock lizard *Agama cyanogaster* (Ruppell) in hot environments. *Ann. Soc. Roy. Zool. Belge 87*, 45-74.

Cuvier, G. L. C. F. D. (1817). *Le Regne Animal, distribue d'apres son organisation pour servir de base a l'histoire naturelle des animaux et d'introduction a l'Anatomie Comparee*. Paris, 4 vols.

Cuvier, G. L. C. F. D. (1829). *Le Regne Animal . . . .* Paris, 5 vols. Edition 2.

Daniel, Paul M. (1961). Notes on the life history of *Agama agama africana* (Hallowell) in Liberia. *Ohio Herpet. Soc. Spec. Publ. 3*, 1-5.

Daudin, F. M. (1802-1803). *Histoire naturelle, generale et particuliere des Reptiles, Ouvrage faisant Suite a l'histoire naturelle generale et particulaire composee par Leclerc de Buffon et redigee de C. S. Sonnini.* Paris. Vols. 1-4, 1802; vols. 5-8, 1803.

Demeter, B. J. (1976). Observations on the care, breeding and behavior of the giant day gecko, *Phelsuma madagascariensis,* at the National Zoological Park, Washington. *Inter. Zoo Yrbk. 16*, 130-133.

Detrait, J. and Boquet P. (1972). Isolement des anticorps and antitoxin α, du venin de *Naja nigricollis* au moyen du Sepharose. *C. r. Hebd. Seanc. Acad. Sci. Paris 274*, 1765-1767.

Detrait, J., Izard, Y. and Boquet, P. (1959). Separation par electrophorese des constituents toxiques de *Naja naja* et de *Naja nigricollis*. *C. r. Seanc. Soc. Biol. 153*, 1722-1724.

Detrait, J., Izard, Y. and Boquet, P. (1960). Relations antigeniques entra un facteur lethal de venin d'*Echis carinatus* et les neurotoxines des venins de *Naja naja* et de *Naja nigricollis*. *C. r. Seanc. Soc. Biol. 154*, 1163-1165.

Diefenbach, C. O. da C. (1973). Integumentary permeability to water in *Caiman crocodilus* and *Crocodylus niloticus* (Crocodilia: Reptilia). *Physio. Zool. 46*, 72-78.

Dixon, James R. and Kroll, James C. (1974). Resurrection of the generic name *Paroedura* for the phyllodactyline geckos of Madagascar, and a description of a new species. *Copeia* 24-30.

Dmi'el, R. (1967). Studies on reproduction, growth and feeding in the snake *Spalerosophis cliffordi* (Colubridae). *Copeia* 332-346.

Dmi'el, R. and Borut, A. (1972). Thermal behavior, heat exchange and metabolism in the desert snake *Spalerosophis cliffordi*. *Physiol. Zool. 45*, 78-94.

Dmi'el, R. and Zilber B. (1971). Water balance in a desert snake. *Copeia* 754-755.

Dollo. L. (1886). Notice sur les Reptiles et Batraciens recueillis par M. le Capitaine Em. Storms dans la region du Tanganyika. *Bull. Mus. R. Hist. Nat. Belgique 4*, 151-160.

Domergue, Ch. A. (1954). Note sur une nouvelle espece ophidienne de Tunisie et d' Afrique du Nord: *Coluber choumowitchi*. *Bull. Soc. Sci. nat. Tunisie 7*, 37-47.

Domergue, Ch. A. (1955). Note sur un serpent nouveau: *Pseudotarbophis gabesi* n. sp. *Bull. Soc. Sci. nat. Tunisie 8*, 119-123.

Domergue, Ch. A. (1959a). Liste des Ophidiens de Tunisue, de l'Algerie et du Maroc. *Arch. Inst. Pasteur Tunis 36*, 157-161.

Domergue, Ch. A. (1959b). Cle de determination des Serpents de Tunisie et Afrique du Nord. *Arch. Inst. Pasteur Tunis 36*, 163-172.

Donndorff, J. A. (1798). *Zoologische Beytrage zur XIII, Ausgabe des Linneischen Natursystems*. Vol. 3, pp. 980. Leipzig.

Doucet, J. (1963). Les serpents de la Republique de Cote d'Ivoire. *Acta trop. 20*, 201-340.

Doumergue, F. (1899). Essai sur la Faune erpetologique de l'Oranie. *Soc. Geogr. Arch. Oran 19*, 197-260, 501-532.

Doumergue, F. (1901). *Essai sur la Faune erpetologique de l'Oranie*. Imp. Typogr. Fouque, Oran. pp. 404. (Reprinted, 1972, Linnaeus Press, Amsterdam).

Dowling, H. G. (1969). Relations of some African colubrid snakes. *Copeia* 234-243.

Dumarey, C. and Boquet, P. (1972). Pouvoir immunogene de la toxine ∝ du venin de *Naja nigricollis* polymerisee par l'aldehyde formique. *C. r. hebd. Seanc. Acad. Sci. 275*, 3053-3055.

Dumeril, A. (1851). Note sur une nouvelle espece de reptile de la famille des Geckotiens, et appartenant au genre Stenodactyle (Stenodactyle queue-cerclee, *Stenodactylus caudicinctus*). *Revue Mag. Zool. (ser. 2) 3*, 479-486.

Dumeril, A. (1853). Prodome de la classification des reptiles ophidiens. *Mem. Acad. Inst. France 23*, 399-536.

Dumeril, A. (1856a). Description des reptiles nouveaux ou imparfaitment connus de la collection de Museum d'Histoire Naturelle, et remarques sur la classification et les caracteres des reptiles. *Arch. Mus. Hist. nat. Paris 8*, 437-588.

Dumeril, A. (1856b). Note sur les Reptiles du Gabon. *Rev. Mag. Zool. (ser. 2) 7*, 369-375, 417-424, 460-470.

Dumeril, A. (1859). Reptiles et poissons de l'Afrique Occidentale. Etude precedee de considerations generales sur leur distribution geographique. *Arch. Mus. Nat. Hist. nat. Paris 10*, 137-268.

Dumeril, A. M. C. and Bibron, G. (1834-1854). *Erpetologie generale ou histoire naturelle complete des reptiles*. Paris, 8 vols.

Dumeril, A. M. C., Bibron, G. and Dumeril, A. H. A. (1851). *Catalogue methodique de la collection des reptiles (Museum d'Histoire naturelle de Paris)*. Paris.

Dumeril, C. (1806). *Analytische Zoologie. Aus dem Franzosischen, mit Zusatzen von L. F. Froriep*. Verlag des Landes — Industrie — Comptoirs, Wiemar.

Dunger, G. T. (1965). *Baikia africana* (family Amphisbaenidae): Ecology and individual variation. *Ann. Mag. Nat. Hist. (ser. 13) 7*, 217-221.

Dunger, G. T. (1966). A new species of the colubrid genus *Mehelya* from Nigeria. *Amer. Mus. Novitates 2268*, 1-8.

Dunger. G. T. (1967a). The lizards and snakes of Nigeria. Part 1. The Chamaeleons of Nigeria. *Nigerian Field 32*, 53-74.

Dunger, G. T. (1967b). The lizards and snakes of Nigeria. Part 2. The Lacertids of Nigeria. *Nigerian Field 32*, 117-131.

Dunger, G. T. (1967c). The lizards and snakes of Nigeria. Part 3. The Monitors and a plated lizard. *Nigerian Field 32*, 170-178.

Dunger, G. T. (1968a). The lizards and snakes of Nigeria. Part 4. The geckos of Nigeria. *Nigerian Field 33*, 18-47.

Dunger, G. T. (1968b). The lizards and snakes of Nigeria. Part 5. The amphisbaenids of Nigeria. *Nigerian Field 33*, 167-192.

Dunger, G. T. (1971a). The snakes of Nigeria. Part 1. The file snakes of Nigeria. *Nigerian Field 36*, 54-71.

Dunger, G. T. (1971b). The snakes of Nigeria. Part 2. The house snakes of Nigeria. *Nigerian Field 36*, 151-163.

Dunger, G. T. (1972a). The snakes of Nigeria. Part 3. The harmless water and marsh snakes of Nigeria. *Nigerian Field 37*, 21-38.

Dunger, G. T. (1972b). The lizards and snakes of Nigeria. Part 6. The skinks of Nigeria. *Nigerian Field 37*, 99-120.

Dunger, G. T. (1973). The lizards and snakes of Nigeria. Part 7. The skinks of Nigeria (continued and completed). *Nigerian Field 38*, 54-80.

Dunson, William A. (1968). Salt gland secretion in the pelagic sea snake *Pelamis. Amer. J. Physiol. 215*, 1512-1517.

Dunson, W. A., Dunson, M. K. and Keith, A. D. (1978). The nasal gland of the montpellier snake *Malpolon monspessulanus:* Fine structure, secretion, composition and a possible role in reduction of dermal water loss. *J. Exp. Zool. 203*, 461-473.

Dunson, W. A. and Ehlert, G. W. (1971). Effects of temperature, salinity, and surface water flow on distribution of the sea snake *Pelamis. Limnol. Oceanogr. 16*, 845-853.

Dupont J. (1977). Catalytic properties of two phospholipases A from cobra (*Naja nigricollis*) venom. *Toxicon 15*, 347-354.

Duvdevani, I. (1972). The anatomy and histology of the nasal cavities and the nasal salt gland in four species of fringe-toed lizards, *Acanthodactylus (Lacertidae). J. Morph. 137*, 353-364.

Duvdevani, I. and Borut, A. (1974a). Mean body temperature and heat absorption in four species of *Acanthodactylus* lizards (Lacertidae). *Herpetologica 30*, 176-181.

Duvdevani, I. and Borut, A. (1974b). Oxygen consumption and evaporative water loss in four species of *Acanthodactylus* (Lacertidae). *Copeia* 155-164.

Duvernoy, M. (1832). Memoire sur les caracteres tires de l'anatomie pour distinguer les serpens venimeux des serpens non venimeux. *Ann. Sci. Nat. 26*, 113-160.

Earl, J. E. and Excell, B. J. (1972). The effects of toxic components of *Naja nivea* (Cape cobra) venom on neuromuscular transmission and muscle membrane permeability. *Comp. Biochem. Physiol. 41A*, 597-615.

Edmondson, Colin R. (1976). Hatching Royal Python (*Python regius*) eggs. *J. S. Western Herpet. Soc. 3*, 14-16.

Eggertsen, Gosta; Fohlman, Jan and Sjoquist, John (1980). *In vitro* studies on complement inactivation by snake venoms. *Toxicon 18*, 87-96.

Eigenberger, F. (1928). Some clinical observations on the action of mamba venom. *Bull. Antiv. Inst. Amer. 2*, 45-46.

Eiselt, J. (1940). Der rassenkreis *Eumeces schneideri* Daudin (Scincidae, Rept.). *Zool. Anz. Leipzig 131*, 209-228.

El-Toubi, M. R. (1938). The osteology of the lizard *Scincus scincus* (Linn.). *Bull. Fac. Sci. Cairo Univ. 14*, 5-38.

El-Toubi, M. R. and Bishai, H. M. (1959). On the anatomy and histology of the alimentary tract of the lizard *Uromastyx aegyptia* (Foskal). *Bull. Fac. Sci. Cairo Univ. 34*, 13-50.

Englebrecht, D. van Z. (1951). Contributions to the cranial morphology of the chamaeleon *Microsaura pumila* Daudin. *Ann. Univ. Stellenbosch 27A*, 1-31.

Essex, R. (1925). Descriptions of two new species of the genus *Acontias* and notes on some other lizards found in the Cape Province. *Rec. Albany Mus. 3*, 332-342.

Ferreira, J. B. (1897). Sobre um *Hemidactylus* novo da Ilha de Anno Bom. *J. Acad. Sci. Lisbon (ser. 2) 4*, 249-251.

Filippi, F. de (1865). *Note di un viaggio in Persia nel 1862*. Milan, 1, 1-396.

Finkeldey, H. (1963). *Python anchietae* Bocage. *Cimbebasia 6,* 23-28.

Fischer, J. G. (1856). Neue Schlangen der Hamburgischen Naturhistorischen Museums. *Abh. Ges. natur. Ver Hamburg 3,* 79-116.

Fischer, J. G. (1884). Uber die von Herrn Dr. G. A. Fischer in Massai-Gebiete (OstAfrika) auf seinerin Veranlassung der geographischen Gesellschaft in Hamburg unternommenen Expedition gesammelten Reptilien, Amphibien und Fische, *Jahrb. Hamburg Wiss. Anst. 1,* 3-32.

Fischer, J. G. (1886). Herpetologische Notizen. *Abh. Ges. Natur. Ver. Hamburg 9,* 1-19.

Fischer, J. G. (1888a). Herpetologische Mitteilungen I. Uber zwei neue Schlangen und einen neuen Laubfrosch aus Kamerun. *Jahrb. Hamburg Wiss. Anst. 5,* 3-10.

Fischer, J. G. (1888b). Uber eine Kollection Reptilien von Angra Pequenna. *Jahrb. Hamburg Wiss. Anst. 5,* 11-17.

Fitzinger, L. J. F. T. (1823). *in* Lichtenstein, H.

Fitzinger, L. J. F. T. (1826). *Neue classification der Reptilien nach ihren naturlichen Verwandtschaften.* Vienna, pp. 66.

Fitzinger, L. J. F. T. (1843). *Systema reptilium. Fasciculus primus, Amblyglossae.* Vienna, pp. 106.

Fitzsimons, V. F. M. (1930). Descriptions of new South African Reptilia and Batrachia, with distribution records of allied species in the Transvaal Museum collection. *Ann. Transvaal Mus. 14,* 20-48.

Fitzsimons, V. F. M. (1932). Preliminary descriptions of new forms of S. African Reptilia and Batrachia from the Vernay-Lang Kalahari Expedition, 1930. *Ann. Transvaal Mus. 15,* 35-40.

Fitzsimons, V. F. M. (1933). Description of five new lizards from the Transvaal and Southern Rhodesia. *Ann. Transvaal Mus. 15,* 273-280.

Fitzsimons, V. F. M. (1937). Three new lizards from South Africa. *Ann. Transvaal Mus. 17,* 275-279.

Fitzsimons, V. F. M. (1938). Transvaal Museum Expedition to South West Africa and Little Namaqualand, May to August 1937. Reptiles and Amphibians. *Ann. Transvaal Mus. 19,* 153-209.

Fitzsimons, V. F. M. (1939). Descriptions of some new species and subspecies of lizards from South Africa. *Ann. Transvaal Mus. 20,* 5-16.

Fitzsimons, V. F. M. (1941). Descriptions of some new lizards from South africa and a frog from Southern Rhodesia. *Ann. Transvaal Mus. 20,* 273-281.

Fitzsimons, V. F. M. (1943). The lizards of South Africa. *Transvaal Mus. Memoir 1,* 1-528.

Fitzsimons, V. F. M. (1946a). An account of the reptiles and amphibians collected on an expedition to the Cape Province, October to December 1940. *Ann Transvaal Mus. 20,* 351-377.

Fitzsimons, V. F. M. (1946b). Notes on some South African snakes, including a description of a new subspecies of *Xenocalamus. Ann. Transvaal Mus. 20,* 379-393.

Fitzsimons, V. F. M. (1948). Notes on some reptiles and amphibians from the Drakensberg, together with a description of a new *Platysaurus* from northern Natal. *Ann. Transvaal Mus. 21,* 73-80.

Fitzsimons, V. F. M. (1959). Some new reptiles from Southern Africa and Southern Angola. *Ann. Transvaal Mus. 23,* 405-409.

Fitzsimons, V. F. M. (1962a). A new worm-snake (*Leptotyphlops*) from South West Africa. *Ann. Transvaal Mus. 24,* 239-240.

Fitzsimons, V. F. M. (1962b). *Snakes of Southern Africa.* Macdonald, London pp. 423.

Fitzsimons, V. F. M. (1966). A checklist, with synoptic keys, to the snakes of Southern Africa. *Ann. Transvaal Mus. 25,* 35-79.

Fitzsimons, V. F. M. (1974). *A field guide to the snakes of Southern Africa.* Collins, 2nd edition. pp. 221.

Fitzsimons, V. F. M. and Brain, C. K. (1958). A short account of the reptiles of the Kalahari Gemsbok National Park. *Koedoe 1,* 99-104.

Fleming, J. (1822). *The Philosophy of Zoology; or a General View of the Structure, Functions, and Classification of Animals.* Constable, Edinburgh *2,* 1-618.

Flower, S. S. (1933). Notes on the recent reptiles and amphibians of Egypt, with a list of the species recorded from that Kingdom. *Proc. zool. soc. London* 735-851.

Fohlman, J. and Eaker, D. (1977). Isolation and characterization of a lethal myotoxic phospholipase A from the venom of the common sea snake (*Enhydrina schistosa*) causing myoglobinuria in mice. *Toxicon 15,* 385.

Forbes, C. D., Turpie, A. G. G., Ferguson, J. C., McNicol, G. P. and Douglas, A. S. (1969). Effect of gaboon viper (*Bitis gabonica*) venom on blood coagulation, platelets, and the fibrinolytic enzyme system. *J. Clinical Pathology 22,* 312.

Forcart, L. (1964). Publication date and holotype of the name *Rachiodon scaber subfasciatus* Muller 1878 (Colubridae, Reptilia). *Copeia 224.*

Forskal, Pehr. (1775). *Descriptiones Animalium, Avium, Amphibiorum, Piscium, Insectorum, Vermium; quae in Itinere orientali observavit Petrus Forskal . . . . post mortem auctoris editit Carsten Niebuhr.* Hauniae pp. 164.

Frair, Wayne (1979). Taxonomic relations among sea turtles elucidated by serological tests. *Herpetologica 35,* 239-244.

Frair, Wayne and Prol, B. (1978). Taxonomic study of the giant leatherback turtle (*Dermochelys coriacea*). *Nat. Geogr. Soc. Res. Rep. 1969,* 187-194.

Frank, G. H. (1951). Contributions to the cranial morphology of *Rhampholeon platyceps* Gunther. *Ann. Univ. Stellenbosch 27A,* 33-67.

Frazzetta, T. H. (1966). Studies on the morphology and function of the skull in the Boidae (Serpentes). Part II. Morphology and function of the jaw apparatus in *Python sebae* and *Python molurus. J. Morph. 118,* 217-296.

Frazzetta, T. H. (1971). Notes upon the jaw musculature of the bolyerine snake, *Casarea dussumieri. J. Herpet. 5,* 61-63.

Frisch, O. V. (1962). Zur biologie des Zwergchamaleons (*Microsaura pumilis*). *Z. Tierpsycol. 19,* 276-289.

Fryklund, L. and Eaker, D. (1973). Complete amino acid sequence of a non-neurotoxic hemolytic protein from the venom of *Hemachatus haemachatus* (African ringhals cobra). *Biochemistry, N. Y. 12,* 661-667.

Fuchs, K. H., Mertens, R. and Wermuth, H. (1974). Zum status von *Crocodylus cataphractus* und *Osteolaemus tetraspis. Stuttgart Beitr. Naturk. 266A,* 1-8.

Fuhn, I. E. (1970). Contribution a la systematique des lygosomines africains (Reptilia, Scincidae). I. Les especes attribuees au genre *Ablepharus. Rev. Roum. Biol.* (*Zool.*) *15,* 379-393.

Gabe, M. (1971). Repartition des cellules histaminergiques dans la paroi gastrique de qualques Reptiles. *C. r. hebd. Seanc. Acad. Sci. Paris* (*ser. D*) *273,* 2287-2289.

Gabe, M. (1972). Donnees histologiques sur les cellules a gastrine des sauropsides. *Archs. hebd. Anat. microsc. Morph. exp. 61,* 175-200.

Gabe, M. and Saint Girons, H. (1965). Contribution is la morphologie comparee du cloaque et des glandes epidermoides de la region cloacale des lepido sauriens. *Mem. Mus. natn. Hist. nat. Paris* (*ser. A*) *33,* 149-292.

Gabe, M. and Saint Girons, H. (1969). Donnees histologiques sur les glandes salivaires des Lepidosauriens. *Mem. Mus. natn. Hist. nat. Paris* (*ser. A*) *58,* 1-112.

Gabe, M. and Saint Girons, H. (1972). A contribution to the histological study of the stomach in Lepidosauria (Reptilia). *Zool. Jb. Anat. 89,* 579-599.

Gaffney, P. J., Marsh, N. A. and Whaler, B. C. (1973). A coagulant enzyme from Gaboon viper venom: Some aspects of its mode of action. *Biochemical Soc. Trans. 1,* 1208.

Galvao, P. E., Tarasantchi, J. and Guertzenstein, P. (1965). Heat production of tropical snakes in relation to body weight and body surface. *Amer. J. Physiol.* *209*, 501-506.

Gans, Carl (1952). The functional morphology of the egg-eating adaptions in the snake genus *Dasypeltis. Zoologica, N. Y. 37*, 209-243.

Gans, Carl (1957). *Dasypeltis medici lamuensis*, a new race of egg-eating snake (Ophidia, Reptilia) from coastal East Africa. *Mus. Comp. Zool. Breviora 79*, 1-13.

Gans, Carl (1959). A taxonomic revision of the African snake genus *Dasypletis* (Reptilia, Serpentes). *Ann. Mus. R. Congo Belge, Terv.* (ser. 8), *Sci. Zool. 74*, 1-237.

Gans, Carl (1960a). The western extremity of the range of *Dasypeltis fasciata*, with notes on ecology and color resemblance. *Copeia* 154-155.

Gans, Carl (1960b). Mimicry in procryptically colored snakes of the genus *Dasypeltis. Evolution 15*, 72-91.

Gans, Carl (1960c). Studies on amphisbaenids (Amphisbaenia, Reptilia). I. A taxonomic revision of the Trogonophinae and a functional inerpretation of the amphisbaenid adaptive pattern. *Bull. Amer. Mus. Nat. Hist. 119*, 129-204.

Gans, Carl (1962). Notes on amphisbaenids (Amphisbaena, Reptilia). I. On the name *Amphisbaena reticulata* Holmer 1787. *Brit. J. Herpet. 3*, 12-13.

Gans, Carl (1967). A checklist of recent amphisbaenians (Amphisbaenia, Reptilia). *Bull. Amer. Mus. Nat. Hist. 135*, 61-106.

Gans, Carl (1969). Amphisbaenians — reptiles specialised for a burrowing existence. *Endeavour 28*, 146-151.

Gans, Carl (1976). Three new spade-snouted amphisbaenians from Angola (Amphisbaenia, Reptilia). *Amer. Mus. Novitates 2590*, 1-11.

Gans, Carl and Broadley, Donald G. (1974). A new dwarfed species of *Monopeltis* from the Middle Zambezi Valley (Reptilia: Amphisbaenia). *Arnoldia Rhodesia* 6(35), 1-5.

Gans, Carl and Hughes, G. M. (1967). The mechanics of lung ventilation in the tortoise *Testudo graeca* Linne. *J. Exp. Biol. 47*, 1-20.

Gans, Carl and Kochva, E. (1966). Notes on amphisbaenids. 22. A systematic review of *Ancylocranium* (Amphisbaenia: Reptilia). *Israel J. Zool. 14*, 87-121.

Gans, Carl; Kraukauer, T. and Paganelli, C. V. (1968). Water loss in snakes: Interspecific and intraspecific variability. *Comp. Biochem. Physiol. 27*, 747-761.

Gans, Carl and Latifi, M. (1971). Redescription and geographical variation of *Monopeltis guentheri* Boulenger (Amphisbaenia, Reptilia). *Amer. Mus. Novitate 2464*, 1-21.

Gans, Carl and Laurent, R. F. (1965). *in* Gans, Carl; Laurent R. F. and Pandit, H. (1965).

Gans, Carl; Laurent, Raymond F. and Pandit, Hemchandra (1965). Notes on a herpetological collection from the Somali Republic. *Ann. Mus. R. Afr. cent.* (ser. 8), *Sci. Zool. 134*, 1-93.

Gans, Carl and Lehman, Grace C. (1973). Studies on amphisbaenians (Amphisbaenia: Reptilia). 5. The species of *Monopeltis* from north of the river Zaire. *Occ. Pap. Mus. Zool., Univ. Michigan 669*, 1-34.

Gans, Carl and Pandit, Hemchandra (1965). *in* Gans, Carl; Laurent, R. F. and Pandit, Hemchandra (1965).

Gans, Carl and Pasteur, G. (1962). Notes on amphisbaenids (Amphisbaenia: Reptilia). 4. On the type locality of *Amphisbaena elegans* Gervais. *Herpetologica 18*, 9-11.

Gans, Carl and Rhodes, C. (1967). *Chirindia* from Tanganyika (Amphisabenia, Reptilia). *Ann. Carnegie Mus. 39*, 1-32.

Gans, Carl and Richmond, N. D. (1957). Warning behavior in snakes of the genus *Dasypeltis Copeia* 269-274.

Gartlan, J. S. and Struhsaker, T. T. (1971). Notes on the habits of the Calabar ground python (*Calabaria reinhardtii* Schlegel) in Cameroon, West Africa. *Brit. J. Herpet. 4,* 201-202.

Gartshore, M. E. (1978). The skink *Mabuya rodengurgi* in Nigeria. *Nigerian Field 43,* 38-41.

Gartside, Don F. and Dessauer, Herbert C. (1977). Immunological evidence on affinities of African *Natrix. Copeia* 190-191.

Gasc, J. P. (1965). Les adaptions anatomiques du lezard apode *Feylinia currori* Gray au fouissage par reptation ondulante. *C. r. hebd. Seanc. Acad. Sci. Paris 260,* 1248-1251.

Gauthier, R. (1956). Note sur trois Agamas du Sahara occidental. *Bull. Soc. Hist. nat. Afrique du Nord 57,* 137-146.

Gauthier, R. (1966). La reproduction chez *Eremias* (*Mesalina*) *rubropunctata* (Licht.) (Lacertidae). Les jeunes et leur croissance. *Bull. I. F. A. N. (ser. A), Sci. Zool. 28,* 1620-1629.

Gawade, S. P. and Bhide, M. B. (1978). Chromatographic seperation of venom of *Enhydrina schistosa* (common sea snake) and characterization of its principal toxic component. *Indian J. Med. Res. 67,* 854-861.

Gaymer, R. (1968). The Indian Ocean giant tortoise *Testudo gigantea* on Aldabra. *J. Zool. London 154,* 341-363.

Geh, S. L. and Toh. H. T. (1978). Ultrastructural changes in skeletal muscle caused by a phospholipase A_2 fraction isolated from the venom of a seasnake, *Enhydrina schistosa. Toxicon 16,* 633-644.

Gene, Josepho, (1839). Synopsis Reptilium Sardiniae indigenorum. *Mem. Reale Acad. Sci. Turin (ser. 2) 1,* 257-286.

Geoffroy Sainte-Hilaire, Etienne C. T. (1827). *Description des Reptiles qui se trouvent en Egypte. In,* Savigny, M. J. C. L., Description de l'Egypte, ou recueil des observations et des recherches qui ont ete faites en Egypte pendant l'Expedition de l'Armee francaise (1798-1801). I. Histoire naturelle. Paris pp. 115-184.

Gervais, P. (1835). Les principaux resultats de l'etude de reptiles envoyes de Barbarie. *Bull. Soc. Sci. nat. France 1835,* 112-114.

Giess, W. (1965). *Chondrodactylus angulifer* (Peters) and *Ptenopus garrulus* (Smith) from the Brandberg. *Cimbebasia 12,* 16-19.

Girard, C. (1853). Description of new species of reptiles collected by the U. S. Exploring Expedition, under the command of Capt. Charles Wilkes, U. S. N. Part II. Including the species of Batrachians exotic to North America. *Proc. Acad. Nat. Sci. Philadelphia 6,* 420-424.

Gistel, J. von N. F. X. (1848). *Naturgeschichte des Thierreichs fur hohere Schulen.* Stuttgart pp. 216.

Gmelin, J. F. (1789-1796). *Caroli a Linne Systema Naturae per Regna tria Naturae, secundum Classes, Ordines, Genera, Species, cum characteribus, differentiis, synonymis, locis. Editio XII, aucta reformata.* Lipsiae, vols 1-3.

Goldberg, Stephen R. and Robinson, Michael D. (1979). Reproduction in two Namib Desert Lacertid lizards (*Aporosaura anchietae* and *Meroles cuneirostris*). *Herpetologica 35,* 169-175.

Goldfuss, G. A. (1820). *Handbuch der Zoologie.* Nurnberg 2, 1-510.

Goodman, John D. and Goodman, Jeanne M. (1976). Possible mimetic behavior of twig-snake, *Thelotornis kirtlandi kirtlandi* (Hallowell). *Herpetologica 32,* 148-150.

Graham, J. B. (1974a). Aquatic respiration in the sea snake *Pelamis platurus*. *Respir. Physiol. 21*, 1-7.

Graham, J. B. (1974b). Body temperature of the sea snake *Pelamis platurus*. *Copeia* 531-533.

Graham, J. B., Gee, J. H. and Robison, F. S. (1975). Hydrostatic and gas exchange functions of the lung of the sea snake *Pelamis platurus*. *Comp. Biochem. Physiol. 50A*, 477-482.

Graham, J. B., Rubinoff, I. and Hecht, M. K. (1971). Temperature physiology of the sea snake *Pelamis platurua*. An index of its colonization potential in the Atlantic Ocean. *Proc. Natl. Acad. Sci. 68*, 1360-1363.

Grandidier, Alfred (1867). Liste des Reptiles nouveaux decouverts, en 1866, sur la Cote sud-ouest de Madagascar. *Revue Mag. Zool. (ser. 2)* 19, 232-234.

Grandison, A. G. C. (1956). On a collection of lizards from West Africa. *Bull. I. F. A. N. 18A*, 224-245.

Grandison, A. G. C. (1961). Preliminary notes on the taxonomy of *Tarentola annularis* and *T. ephippiata* (Sauria: Gekkonidae). *Zoologische Mededelingen 38*, 1-14.

Grandison, A. G. C. (1968). Nigerian lizards of the genus *Agama* (Sauria: Agamidae). *Bull. Brit. Mus. Nat. Hist. (Zool.)* 17, 65-90.

Grasset, E. and Schaafsma, A. W. (1940a). Studies on the venom of the boomslang (*Dispholidus typus*). *S. Afr. Med. J. 14*, 236-241.

Grasset, E. and Schaafsma, A. W. (1940b). Antigenic characteristics of Boomslang (*Dispholidus typus*) venom and preparation of a specific antivenene by means of formalized venom. *S. Afr. Med. J. 14*, 484-489.

Grasset, E. and Zoutendyk, Z. (1938). Studies on the Gaboon viper (*Bitis gabonica*) and the preparation of a specific therapeutic antivenene. *Trans. R. Soc. Trop. Med. Hyg. 31*, 445-450.

Gray, J. E. (1825). A synopsis of the genera of reptiles and amphibians, with a description of some new species. *Ann. Philos. Soc. London (ser. 2) 10*, 193-217.

Gray, J. E. (1827). A synopsis of the genera of the Saurian Reptiles in which some new Genera are indicated, and the others reviewed by actual Examination. *Philos. Mag. (ser. 2) 3*, 53-56.

Gray, J. E. (1828). *Spicilegia Zoologica: or original figures and short systematic descriptions of new and unfigured animals.* London. Part I. and II. pp. 12 and 11 plates.

Gray, J. E. (1831a). A synopsis of the species of Class Reptilia. In (sepearately paginated appendix to) Griffith, E. and Pidgeon, E.: *The animal kingdom arranged in conformity with its organisation by the Baron Cuvier with additional descriptions of all the species hither named, and of many before noticed.* Whittaker, Treacher and Co., London 9, 1-110.

Gray, J. E. (1831b). Description of a new species of chamaeleon discovered by Capt. Owen in Africa. *Zool. Misc.* 7.

Gray, J. E. (1838a). A new lizard *Phyllodactylus lineatus*, and new land shells from South Africa. *in* Alexander '*An Expedition into the interior of South Africa'. 2*, 268.

Gray, J. E. (1838b). Catalogue of the slender-tongued saurians, with descriptions of many new genera and species. *Ann. Mag. Nat. Hist. (ser. 1) 1*, 274-283.

Gray, J. E. (1839). Catalogue of the slender-tongued saurians, with descriptions of many new genera and species. *Ann. Mag. Nat. Hist. 2*, 287-293, 331-337.

Gray, J. E. (1842a). Synopsis of the species of prehensile-tailed snakes, or family Boidae. *Zool. Misc.* 41-46.

Gray, J. E. (1842b). Descriptions of some new species of Reptiles, chiefly from the British Museum Collection. *Zool. Misc.* 57-59.

Gray, J. E. (1842b). Monographic Synopsis of the Vipers, or the family Viperidae. *Zool. Misc.* 68-71.

Gray, J. E. (1844). *Catalogue of the tortoises, crocodiles, and amphisbaenians, in the collection of the British Museum.* London pp. 80.

Gray, J. E. (1845). *Catalogue of the specimens of lizards in the British Museum.* London pp. 289.

Gray, J. E. (1849). *Catalogue of the specimens of snakes in the collection of the British Museum.* London pp. 125.

Gray, J. E. (1858). Description of a new genus of Boidae from Old Calabar and a list of W. African Reptiles. *Proc. zool. soc. London* 154-167.

Gray, J. E. (1864a). Notes on some new lizards from Southeastern Africa, with descriptions of several new species. *Proc. zool. soc. London* 58-63.

Gray, J. E. (1864b). Revision of the genera and species of Chamaeleontidae, with description of some new species. *Proc. zool. soc. London* 465-479.

Gray, J. E. (1865a). A revision of the genera and species of amphisbaenians with the descriptions of some new species now in the collection of the British Museum. *Proc. zool. soc. London* 442-455.

Gray, J. E. (1865b). Descriptions of two new genera of lizards from Damaraland. *Proc. zool. soc. London* 640-642.

Gray, J. E. (1869). Notes on the families and genera of tortoises (Testudinata) and on the characters afforded by the study of their skulls. *Proc. zool. soc. London* 165-225.

Greene, Harry W. (1973). Defensive tail display by snakes and amphisbaenians. *J. Herpet. 7*, 143-161.

Greenham, R. (1978). Spitting cobra (*Naja mossambica pallida*) bite in a Kenyan child. *Trans. Roy. Soc. Trop. Med. Hyg. 72*, 674.

Greer, A. E. (1967). The generic relationships of the African genus *Eumecia*. *Breviora 276*, 1-9.

Greer, A. E. (1970a). A subfamilial classification of scincid lizards. *Bull. Mus. Comp. Zool. 139*, 151-183.

Greer, A. E. (1970b). The systematics and evolution of the Subsaharan Africa, Seychelles, and Mauritius scincine scincid lizards. *Bull. Mus. Comp. Zool. 140*, 1-23.

Greer, A. E. (1976). On the evolution of the giant Cape Verde scincid lizard *Macroscincus coctei. J. Nat. Hist. 10*, 691-712.

Greer, A. E. (1977). The systematics and evolutionary relationships of the scincid lizard genus *Lygosoma. J. Nat. Hist. 11*, 515-540.

Gregoire, J. and Rochat, H. (1977). Amino acid sequences of neurotoxins I and III of the Elapidae snake *Naja mossambica mossambica*. *Eur. J. Biochem. 80*, 283-293.

Grenot, C. (1967). Observations physico-ecologiques sur thermoregulation dans le lezard agamid *Uromastix acanthinurus,* Bell. *Bull. soc. zool. France 92*, 51-66.

Grenot, C. (1968a). Etude comparative de la resistance a la chaleur d'*Uromastyx acanthinurus* et de *Varanus griseus. Terre et la Vie 4*, 390-409.

Grenot, C. (1968b). Sur l'excretion nasale de scls chez le lezard saharien: *Uromastyx acanthinurus. C. r. Acad. Sci. Paris D266*, 1871-1874.

Grenot, C. (1974). Polymorphisme chromatique du lezard Agamide *Uromastyx acanthinurus* Bell dans les populations du Sahara Nord Occidental. *Bull. soc. zool. France 99*, 153-164.

Groves, F. (1973). Reproduction and venom in Blanding's tree snake, *Boiga blandingi. Inter. Zoo Yrbk. 13*, 106-108.

Grunwald, E. (1931). La torsion intestinale chez les reptiles. *Archs. Anat. Histol. Embryol. 14*, 167-203.

Grys, P. de (1938). *Gerrhosaurus maltzahni* spec. nov. *Zool. Anz. Leipzig 124*, 58-60.

Guggisberg, C. A. W. (1972). *Crocodiles. Their natural history, folklore and conservation.* David and Charles pp. 204.

Guibe, J. (1948). Sur le dimorphisme sexuel des especes du genre *Langaha* (Ophidien). *C. r. Acad. Sci. Paris* 1219-1220.

Guibe, J. (1949). Revision du genre *Langaha*. Le dimorphisme sexuel et ses consequences taxonomiques. *Mem. Inst. Sci. Madagascar A3*, 95-105.

Guibe, J. (1954). Etude de *Liopholidophis lateralis* (D. et B.) et description d'une espece nouvelle (Reptilia). *Mem. Inst. Sci. Madagascar A9*, 241-246.

Guibe, J. (1956). Revision des especes malagaches du genre *Phyllodactylus* Gray *Mem. Inst. Sci. Madagascar A10*, 245-250.

Guibe, J. (1958). Les serpents de Madagascar. *Mem. Inst. Sci. Madagascar A12*, 189-260.

Guichenot, Alphonse (1850). Histoire naturelle des Reptiles et des Poissons. In, *Exploration scientifique de l'Algerie pendant les Annees 1840, 1841, 1842, Sci. Phys. Zool.* Paris 5, 1-144.

Guillin, M. C., Bezeaud, A. and Menach D. (1978). The mechanism of activation of human prothrombin by an activator isolated from *Dispholidus typus* venom. *Biochim. Biophys. Acta 573*, 160-168.

Gunther, A. (1858). *Catalogue of the colubrine snakes in the collection of the British Museum.* London pp. 281.

Gunther, A. (1859). Description of a new genus of West African snakes and revision of the South American *Elaps. Ann. Mag. Nat. Hist.* (ser. 3) 4, 161-174.

Gunther, A. (1862). On new species of snakes in the collection of the British Museum. *Ann. Mag. Nat. Hist.* (ser. 3) 9, 124-132.

Gunther, A. (1863a). On some species of tree-snakes (*Ahaetulla*). *Ann. Mag. Nat. Hist.* (ser. 3) 11, 283-287.

Gunther, A. (1863b). Third account of new species of snakes in the collection of the British Museum. *Ann. Mag. Nat. Hist.* (ser. 3) 12, 348-365.

Gunther, A. (1864). Report on a collection of reptiles and fishes made by Dr. Kirk in the Zambesi and Nyassa regions. *Proc. zool. soc. London* 303-314.

Gunther, A. (1865). Fourth account of new species of snakes in the collection of the British Museum. *Ann. Mag. Nat. Hist.* (ser. 3) 15, 89-98.

Gunther, A. (1866). Fifth account of new species of snakes in the collection of the British Museum. *Ann. Mag. Nat. Hist.* (ser. 3) 18, 24-29.

Gunther, A. (1868). Sixth account of new species of snakes in the collection of the British Museum, *Ann. Mag. Nat. Hist.* (ser. 4) 1, 413-429.

Gunther, A. (1872a). Seventh account of new species of snakes in the collection of the British Museum. *Ann. Mag. Nat. Hist.* (ser. 4) 9, 15-37.

Gunther, A. (1872b). Description of three new species of *Eremias. Ann. Mag. Nat. Hist.* (ser. 4) 9, 381-382.

Gunther, A. (1873). Notes on and descriptions of some lizards with rudimentary limbs in the British Museum. *Ann. Mag. Nat. Hist.* (ser. 4) 12, 145-148.

Gunther, A. (1874). Description of some new or imperfectly known species of Reptiles from the Cameroon Mountains. *Proc. zool. soc. London* 442-445.

Gunther, A. (1877). Descriptions of some new species of reptiles from Madagascar. *Ann. mag. Nat. Hist.* (ser. 4) 19, 313-319.

Gunther, A. (1879a). On mammals and reptiles from Johanna, Comoro Islands. *Ann. Mag. Nat. Hist.* (ser. 5) 3, 215-219.

Gunther, A. (1879b). Description of a new species of chamaeleon from Madagascar. *Ann. Mag. Nat. Hist.* (ser. 5) 4, 246-247.

Gunther, A. (1879c). Descriptions of four new species of chamaeleon from Madagascar. *Proc. zool. soc. London* 148-150.

Gunther, A. (1880). Description of new species of reptiles from eastern Africa. *Ann. Mag. Nat. Hist.* (ser. 5) 6, 234-238.

Gunther, A. (1881a). In, Oates, F., *Matabeleland and the Victoria Falls, a naturalist's wanderings in the interior of South Africa.* London pp. 383.

Gunther, A. (1881b). Seventh contribution to the knowledge of the fauna of Madagascar. *Ann. Mag. Nat. Hist.* (ser. 5) 7, 358-359.

Gunther, A. (1888). Contribution to the knowledge of snakes of tropical Africa. *Ann. Mag. Nat. Hist.* (ser. 6) 1, 322-335.

Gunther, A. (1890). Tenth contribution to the knowledge of the fauna of Madagascar. *Ann. Mag. Nat. Hist.* (ser. 6) 5, 71-74.

Gunther, A. (1891). Eleventh contribution to the knowledge of the fauna of Madagascar. *Ann. Mag. Nat. Hist.* (ser. 6) 8, 287.

Gunther, A. (1893). Report on a collection of reptiles and batrachians transmitted by Mr. H. H. Johnston, C. B., from Nyassaland. *Proc. zool. soc. London* (for 1892) 555-558.

Gunther, A. (1894). Report on the collection of Reptiles and Fishes made by Dr. J. W. Gregory during his Expedition to Mount Kenia. *Proc. zool. soc. London* 84-88.

Gunther. A. (1895). Notice of reptiles and batrachians collected in the eastern half of tropical Africa. *Ann. Mag. Nat. Hist.* (ser. 6) 15, 523-529.

Gunther, A. (1903). Reptiles from Rio de Oro, Western Sahara. *Nov. Zool. Tring 10,* 298-299.

Gygax, P. (1971). Entwicklung, Bau und Kunktion der Giftdruse (Duvernoy's gland) von *Natrix tessellata. Acta tropica 28,* 225-274.

Haacke, W. D. (1964). Description of two new species of lizards and notes on *Fitzsimonsia brevipes* (Fitzsimons) from the central Namib Desert. *Sci. Pap. Namib Desert Res. Stat. 25,* 1-15.

Haacke, W. D. (1965). Additional Notes on the Herpetology of South West Africa with descriptions of two new Subspecies of Geckos. *Cimbebasia 11,* 1-40.

Haacke, W. D. (1975a). Description of a new adder (Viperidae, Reptilia) from Southern Africa, with a discussion of related forms. *Cimbebasia 4A,* 115-128.

Haacke, W. D. (1975b). The burrowing geckos of Southern Africa, I. (Reptilia: Gekkonidae). *Ann. Transvaal Mus. 29,* 197-243.

Haacke, W. D. (1976a). The burrowing geckos of Southern Africa, 2. (Reptilia: Gekkonidae). *Ann. Transvaal Mus. 30,* 13-28.

Haacke, W. D. (1976b). The burrowing geckos of Southern Africa, 3. (Repilia: Gekkonidae). Genus *Colupus* Peters. *Ann. Transvaal Mus. 30,* 29-39.

Haacke, W. D. (1976c). The burrowing geckos of Southern Africa, 4. (Reptilia: Gekkonidae). Genus *Chondrodactylus* Peters. *Ann. Transvaal Mus. 30,* 53-70.

Haacke, W. D. (1976d). The burrowing geckos of Southern Africa, 5. (Reptilia: Gekkonidae). Phylogenetic and taxonomic affinities. *Ann. Transvaal Mus. 30,* 71-89.

Haacke, W. D. and Burton M. N. (1978). On two little known snakes from the tropical subtraction zone of southeastern Africa. *Ann. Transvaal Mus. 31,* 43-50.

Haas, G. (1930). Uber die Kaumuskulatur und die Schadelmechanik einiger Wuhlschlangen. *Zool. Jb., Abt. Anat. 52,* 95-218.

Haas, G. (1931a) Die Kiefermuskulatur und die Schadelmechanik de Schlangen in vergleichender Darstellung. *Zool. Jb., Abt. Anat. 53,* 127-198.

Haas, G. (1931b). Uber die Morphologie der Kiefermuskulatur und die Schadelmechanik einiger Schlangen. *Zool. Jb., Abt. Anat. 54,* 333-416.

Haas, G. (1932). Untersuchungen uber den Kieferapparat und die verwandtschaft-lichen Zusammenhange der Schlangen. *Forsch. Fortschr. 8*, 207-210.

Haas, G. (1937). The structure of the nasal cavity in *Chamaeleo chamaeleon* (Linnaeus). *J. Morphol. 61*, 433-451.

Haas, G. (1938). A note on the origin of solenoglyph snakes. *Copeia* 73-78.

Haas, G. (1951). Remarks on the status of the lizard *Eremias olivieri*. *Copeia* 274-276.

Haas, G. (1952). The head muscles of the genus *(Causus)* (Ophidia, Solenoglypha) and some remarks on the origin of the Solenoglypha. *Proc. zool. soc. London 122*, 573-592.

Haas, G. (1973). Muscles of the jaws and associated structures in the Rhynchocephalia and Squamata. In, Gans, C. and Parsons, T. S. (editors), *Biology of the Reptilia 4*, 285-490.

Hager, P. K. (1905). Die Kiefermuskeln der Schlangen und ihre Beziehungen zu den Speicheldrusen. *Zool. Jb., Abt. Anat. 22*, 173-224.

Hahn, Donald E. (1978a). Liste der rezenten Amphibien und Reptilien. Scolecophidia: Anomalepidae, Leptotyphlopidae, und Typhlopidae. *Des Tierreich 101*, 1-65.

Hahn, Donald E. (1978b). A brief review of the genus *Leptotyphlops* (Reptilia, Serpentes, Leptotyphlopidae) of Asia, with description of a new species. *J. Herpet. 12*, 477-489.

Hahn, Donald E. (1979). A new species of *Cynisca* (Amphisbaenidae) from the Ivory Coast. *Copeia* 122-125.

Hallowell, E. (1842). Description of a new genus of Serpents from Western Africa. *J. Acad. Sci. Philadelphia 8*, 336-338.

Hallowell, E. (1844). Description of new species of African Reptiles. *Proc. Acad. Nat. Sci. Philadelphia* 169-172.

Hallowell, E. (1852a). Description of new species of Reptilia from western Africa. *Proc. Acad. Nat. Sci. Philadelphia* 62-65.

Hallowell, E. (1852b). On a new genus and two new species of African snakes. *Proc. Acad. Nat. Sci. Philadelphia* 203-205.

Hallowell, E. (1854a). Remarks on the geographical distribution of Reptiles with descriptions of several species supposed to be new and correction of former paper. *Proc. Acad. Nat. Sci. Philadelphia* 98-105.

Hallowell, E. (1854b). Descriptions of new reptiles from Guinea. *Proc. Acad. Nat. Sci. Philadelphia* 193-194.

Hallowell, E. (1857). Notice of a collection of reptiles from the Gaboon Country, West Africa, recently presented to the Academy of Natural Sciences of Philadelphia, by Dr. Henry A. Ford. *Proc. Acad. Nat. Sci. Philadelphia* 48-72.

Halstead, L. B. (1970). Some observations of the rainbow lizard *Agama agama* at the University of Ife. *Nigerian Field 35*, 86-89.

Hamilton III, William J. and Coetzee, Cornelius G. (1969). Thermoregulatory behaviour of the vegetarian lizard *Angolosaurus skoogi* on the vegetationless northern Namib Desert dunes. *Sci. Pap. Namib Desert Res. Stat. 47*, 95-103.

Hara, K. and Kikuchi, F. (1978). Breeding the west African dwarf crocodile *Osteolaemus tetraspis tetraspis* at Ueno Zoo, Tokyo, *Inter. Zoo Yrbk. 18*, 84-87.

Harding, K. A. and Welch, K. R. G. (1980). *Venomous Snakes of the World. A Checklist*. Pergamon Press pp. 188.

Harris, V. A. (1964). *The Life of the rainbow lizard*. Hutchinson, London pp. 174.

Hartert, Ernst (1913). Expedition to the Central Western Sahara. V. Reptiles Batrachians. *Novi. Zool. Tring 20*, 76-84.

Hassan, F. and El Hawary, M. F. S. (1977). Fractionation of the snake venoms of *Cerastes cerastes* and *Cerastes vipera*. *Toxicon 15*, 170-173.

Hawes, B. (1974). Observations of the feeding habits of the Royal Python, *Python regius. J. South West. Herpet. Soc. 2,* 5-6.

Heatwole, H. and Davison, E. (1976). A review of caudal luring in snakes with notes on its occurence in the Saharan sand viper, *Cerastes vipera. Herpetologica 32,* 332-336.

Hecht, M. K., Kropach, C. and Hecht, B. M. (1974). Distribution of the yellow-bellied sea snake *Pelamis platurus,* and its significance in relation to the fossil record. *Herpetologica 30,* 387-395.

Hediger, H. (1928). Die Tierwett auf einer marokkanischen Farm. *Blatt. Aquar. Terrar. Kunde 39,* 406-408.

Hediger, H. (1935). Herpetologische Beobachtungen in Marokko. *Verhandl. Naturf. Ges. Basel 46,* 1-49.

Hediger, H. (1937). Herpetologische Beobachtungen in Marokko, II. Zur Herpetofauna der Umgebung von Ouezzan (und Tanger). *Verhandl. Naturf. Ges. Basel 48,* 183-192.

Hellmich, W. (1957). Die Reptilienausbeute der Hambergischen Angola-Expedition. *Mitt. Hamberg zool. Mus. Inst. 55,* 39-80.

Hendrickson, John R. (1958). The green turtle, *Chelonia mydas* (Linn.) in Malaya and Sarawak. *Proc. zool. soc. London 130,* 455-535.

Herlant, M. and Pasteels, J. (1955). Etude comparee du developpement de l'hypophyse chez deux lacertiliens africains: *Mabuia megalura* (Peters) et *Chamaeleo bitaeniatus ellioti* (Gunther). *Arch. Biol., Stockholm 66,* 167-193.

Hermann, Johannes (1804). *Observations zoologicae quibus novae complures aliaeque animalium species describuntur et illustrantur (opus posthumum), edidit Fridericus Ludovicus Hammer.* Paris pp. 332.

Hewitt, J. (1909). Description of a new species of *Platysaurus* and notes on the specific characters of certain species of Zonuridae, together with synoptical keys to all South African species, and a resume of our knowledge of their distribution; and a key to all known genera of South African lizards. *Ann. Transvaal Mus. 2,* 29-40.

Hewitt, J. (1915). Descriptions of two new South African lizards, *Tetradactylus laevicauda* and *T. fitzsimonsi. Ann. Transvaal Mus. 2,* 29-40.

Hewitt, J. (1923). Descriptions of two new South African geckos of the genus *Pachydactylus. Ann. Natal Mus. 5,* 67-71.

Hewitt, J. (1925). On some new species of Reptiles and Amphibians from South Africa. *Rec. Albany Mus. 3,* 343-368.

Hewitt, J. (1926a). Descriptions of new and little known lizards and batrachians from South Africa. *Ann. S. Africa Mus. 20,* 413-431.

Hewitt, J. (1926b). Some new or little known reptiles and batrachians from South Africa. *Ann. S. Africa Mus. 20,* 473-490.

Hewitt, J. (1926c). Descriptions of some new species of batrachians and lizards from South Africa. *Ann. Natal Mus. 5,* 435-448.

Hewitt, J. (1927). Further descriptions of reptiles and batrachians from South Africa. *Rec. Albany Mus. 3,* 371-415.

Hewitt, J. (1929). On some Scincidae from South Africa, Madagascar and Ceylon. *Ann. Transvaal Mus. 13,* 1-8.

Hewitt, J. (1932). On some new species and subspecies of South African batrachians and lizards. *Ann. Natal Mus. 7,* 105-128.

Hewitt, J. (1933). Descriptions of some new reptiles and a frog from Rhodesia. *Occ. Pap. Rhodesia Mus. 2,* 45-50.

Hewitt, J. (1935). Some new forms of batrachians and reptiles from South Africa. *Rec. Albany Mus. 4,* 283-357.

Hewitt, J. (1937a). Descriptions of South African lizards. *Ann. Natal Mus. 8,* 199-209.

Hewitt, J. (1937b). *A guide to the vertebrate fauna of the Eastern Cape Province, South Africa. Part II. Reptiles, Amphibians and freshwater Fishes.* Grahamstown pp. 141.

Hewitt, J. (1938). Descriptions of new forms of the genus *Acontias* Linn. *Trans. Roy. Soc. S. Africa 26,* 39-48.

Heyden, C. H. G. von (1827). Reptilien. In, Ruppell, E.: *Atlas zu Reise im nordlichen Afrika. I. Zoologie.* Frankfurt pp. 1-24.

Heyder, G. (1968a). Das Respirationssystem von *Typhlops vermicularis* Merrem (1820), unter besonderer Berucksichtigung der Tracheallunge. *Zeits. Wiss. Zool. 117,* 392-402.

Heyder, G. (1968b). Das Urogenitalsystem von *Typhlops vermicularis* Merrem (1820). *Morph. Jahrbuch. 112,* 594-605.

Heyder, G. (1973). Das Blutgefassystem von *Typhlops vermicularis* Merrem (1820). *Morph. Jahrbuch. 119,* 492-513.

Hibbard, E. and Lavergne, J. (1972). Morphology at the retina of the seasnake *Pelamis platurus. J. Anat. 112,* 125-136.

Hiestand, P. C. and Hiestand, R. R. (1979). *Dispholidus typus* (Boomslang) snake venom: Purification and properties of the coagulant principle. *Toxicon 17,* 489-498.

Hillenius, D. (1959). The differentiation within the genus *Chamaeleo* Laurenti 1768. *Beaufortia 8,* 1-92.

Hillenius, D. (1963). Notes on chamaeleons. I. Comparative cytology: aid and new complications in chamaeleon taxonomy. *Beaufortia 9,* 201-218.

Hiller, U. (1977). Regeneration and degeneraion of setae-bearing sensilla in scales of gekkonid lizard *Tarentola mauritanica* L. *Zool. Anz. 199,* 113-120.

Hirji, K. N. (1977). The occurrence of Pentastomids in the lungs of *Mabuya striata* (Reptilia: Scincidae). *Dar es Salaam Univ. Sci. J. 3,* 39-40.

Hirth, H. F. (1971). Synopsis of biological data on the green turtle *Chelonia mydas* (Linnaeus) 1758. *F. A. O. Fish Synop. 85.*

Hirth, H. F. and Latif, E. M. Abdel (1979). Deep body temperatures of the Nile monitor (*Varanus niloticus*) taken by radio telemetry. *J. Herpet. 13,* 367-368.

Hokama, Y., Iwanaga, S., Tatsuki, T. and Suzuki, T. (1976). Snake venom proteinase inhibitors—III. Isolation of five polypeptide inhibitors from the venom of *Hemachatus haemachatus* (ringhals cobra) and *Naja nivea* (Cape cobra) and the complete amino acid sequence of two of them. *J. Biochem. (Tokyo) 79,* 559.

Honegger, R. E. (1967). Beobachtungen und den Riesenchildkroten (*Testudo gigantea* Schweigger) der Inseln im indischen Ozean. *Salamandra 3,* 101-121.

Hoofien, J. H. (1962). An unusual congregation of the gekkonid lizard *Tarentola annularis* (Geoffroy). *Herpetologica 18,* 54-56.

Hoogmoed, M. S. (1973a). Herpetologische waarnemingen in Ghana (2). De skinken I. *Het Aquarium 43,* 258-264.

Hoogmoed, M. S. (1937b). Herpetologische Beobachtungen in Ghana (II). Die Skinke. *D. A. T. Z. 26,* 217-222.

Hoogmoed, M. S. (1974). Ghanese lizards of the genus *Mabuya* (Scincidae, Sauria, Reptilia). *Zool. Verhandelingen 138,* 1-62.

Horn, Haja (1947). The embryonic development of the pituitary body in the chamaeleon. *Copeia* 262-268.

Horton, D. R. (1972). A new scincid genus from Angola. *J. Herpet. 6,* 17-20.

Howard, N. L. (1975). Phospholipase A_2 from puff adder (*Bitis arietans*) venom. *Toxicon 13,* 21-30.

Howell, K. M. and Mng'ong'o, G. (1978). *Lygodactylus picturatus* feeding on *Trigona* sp. honey. *E. Afr. Nat. Hist. Soc. Bull.* 133-134.

Huey, R. B. and Pianka, E. R. (1977). Seasonal variation in thermoregulatory behavior and body temperature of diurnal Kalahari lizards. *Ecology 58*, 1066-1075.

Huey, R. B., Pianka, E. R., Egan, M. E. and Coons, L. W. (1974). Ecological shifts in sympatry: Kalahari fossorial lizards (*Typhlosaurus*). *Ecology 55*, 304-316.

Huffman, T. N. (1974). Reproduction of a gaboon viper *Bitis gabonica gabonica* in captivity. *Arnoldia Rhodesia 6*, 1-7.

Hughes, B. (1976). Zoogeography of West African False cobras (*Pseudohaje* spp.). *Bull. I. F. A. N. A38*, 457-466.

Hughes, B. (1978). Latitudinal clines and ecogeography of the West African night adder *Causus maculatus* (Hallowell, 1842). Serpentes, Viperidae. *Bull. I. F. A. N. A39*, 358-384.

Hughes, B. and Barry, D. H. (1969). The snakes of Ghana: a checklist and key. *Bull. I. F. A. N. A31*, 1004-1041.

Hughes, G. R., Bass, A. J. and Mentis, M. T. (1968). New African record for Ridley turtle (*Lepidochelys olivacea*). *Copeia 423*.

Hulselmans, J. L. J., Roo, A. de and Vree, F. de (1970). Contribution a l'herpetologie de la Republique du Togo. 1. Liste preliminaire des Serpents recoltes par la premiere Mission Zoologique belge au Togo. *Revue Zool. Bot. Afr. 81*, 193-196.

Hulselmans, J. L. J., Roo, A. de and Vree, F. de (1971). Contribution a l'herpetologie de la Republique du Togo. 3. Liste preliminaire des Serpents recoltes par la troisieme Mission Zoologique belge au Togo. *Revue Zool. Bot. Afr. 83*, 47-49.

Hurwitz, B. J. and Hull, P. R. (1971). Berg-adder bite. *S. Afr. Med. J. 45*, 969-971.

Hussein, M. F. (1961a). Studies on temperature relationships of Egyptian desert reptiles. IV. On the heat death of the snake *Zamenis diadema. Bull. Fac. Sci. Cairo Univ. 37*, 33-42.

Hussein, M. F. (1961b). Studies on the temperature relation of Egyptian desert reptiles. V. On the body temperature of the snake *Zamenis diadema. Bull. Fac. Sci. Cairo Univ. 37*, 59-65.

Hussein. M. F. (1965). On the temperature preferendum of the snake *Zamenis diadema. Bull. Fac. Sci. Cairo Univ. 39*, 169-177.

Inger, R. F. (1948). The systematic status of the crocodile *Osteoblepharon osborni. Copeia* 15-19.

Inger, R. F. and Clark, P. J. (1943). Partition of the genus *Coluber. Copeia* 141-145.

Ionides, C. J. P. and Pitman, C. R. S. (1965). Notes on two East African venomous snakes, *Echis carinatus pyramidum* (Geoffroy), Egyptian saw-scaled viper and *Vipera hindii* Boulenger, Montane viper. *J. E. Afr. Nat. Hist. Soc. 15*, 116-121.

Isemonger, R. M. (1962). *Snakes of Africa, southern, central and east.* Nelson pp. 236.

Islam, A. (1956). The postcranial skeleton of *Uromastix hardwickii. Biol. Lahore 2*, 231-246.

Issekutz, B. von and Vegh, F. (1928). Beitrage zur Wirkung des Insulins. III. Mitteilung Wirkung auf den Gasstoffwechsel der Schildkrote. *Biochem. Z. 192*, 383-389.

Izard, Y. and Boquet, P. (1958). Recherches sur les venins de *Vipera xanthina palaestinae* et d'*Echis carinata. Ann. Inst. Pasteur, Paris 94*, 583-589.

Jacobsen, N. H. G. (1972). Some notes on the biology and behavior of the Transvaal girdled lizard (*Cordylus vittifer*) in captivity. *J. Herpet. Ass. Afr. 9*, 35-37.

Jacobshagen, E. (1920). Zur Morphologie des Oberflachenreliefs der Rumpfdarmschleimhaut der Reptilien. *Jena Z. Naturw. 56*, 361-430.

James, F. C. and Porter, W. P. (1979). Behavior-microclimate relationships in the African Rainbow lizard, *Agama agama. Copeia* 585-593.

Jan, G. (1858). Plan d'une iconographie descriptive des ophidiens, et description sommaire de nouvelles especes de serpents. *Revue Mag. Zool.* 9, 438-449 and 514-527.

Jan. G. (1859). Additions et rectifications aux Plan et Prodome de l'Iconographie descriptive des Ophidiens. *Revue Mag. Zool.* 11, 503-512.

Jan. G. (1860). *Iconographie generale des Ophidiens.* Milan and Paris. 3 vols. pp. 100.

Jan. G. (1861). Note sulla famiglia die tiflopidi sui loro generi e sulle specie del genere *Stenostoma. Archo. zool. Anat. Fis.* 1, 178-199.

Jan. G. (1862a). Enumerazione sisttematica della specie d'ofidi del gruppo *Calamaria. Arch. zool. Anat. phys. Paris* 2, 1-76.

Jan. G. (1862b). Enumeratio sistematica degli Ofidi appartimente al Gruppo Coronellidae. *Arch. zool. Anat. phys. Paris* 2, 211-230.

Jan. G. (1863). *Elenco Sistematico Degli Ofidi Descritti e Disegnati per l'Iconografia Generale.* Milan pp. 143.

Janecek, J. (1976). An exceptionally large puff adder brood, *Bitis arietans. Inter. Zoo Yrbk.* 16, 85-86.

Joubert, F. J. (1977a). Snake venom toxin. The amino acid sequence of three toxins (CM-2h, CM-4b and CM-6) from *Naja haje annulifera* (Egyptian cobra) venom *Hoppe-Seyler's Z. physiol. Chem.* 358, 79-96.

Joubert, F. J. (1977b). Snake venom toxins. The amino acid sequence of two toxins (CM-2a and CM-3) from *Naja haje annulifera* (Egyptian cobra) venom. *Hoppe-Seyler's Z. physiol. Chem.* 358, 377-390.

Joubert, F. J. (1977c). *Naja mossambica mossambica* venom. Purification, some properties and the amino acid sequence of three phospholipases A (CM-1, CM-II and CM-III). *Biochem. biophys. Acta 360,* 156.

Joubert, F. J., Strydom, A. J. C. and Taljaard, N. (1978). Snake venoms. The amino-acid sequence of protein S_5C_4 from *Dendroaspis jamesoni kaimosae* (Jameson's mamba) venom. *Hoppe-Seyler's Z. physiol. Chem.* 359, 741-749.

Joubert, F. J. and Strydom, D. J. (1978). Snake venoms. The amino-acid sequence of trypsin inhibitor E of *Dendroaspis polylepis polylepis* (black mamba) venom. *Eur. J. Biochem.* 87, 191-198.

Joubert, F. J. and Taljaard, N. (1978a). *Naja haje haje* (Egyptian cobra) venom. Purification, some properties and the amino acid sequences of four toxins (CM-7, CM-8, CM-9 and CM-10b). *Biochim. biophys. Acta 534,* 331-340.

Joubert, F. J. and Taljaard, N. (1978b). Purification, some properties and the primary structures of three reduced and S-carboxymethylated toxins (CM-5, CM-6 and CM-10a) from *Naja haje haje* (Egyptian cobra) venom. *Biochim. biophys. Acta 573,* 1-8.

Joubert, F. J. and Taljaard, N. (1978c). *Naja haje haje* (Egyptian cobra) venom, some properties and the complete primary structure of three toxins (CM-2, CM-11 and CM-12). *Eur. J. Biochem.* 90, 359-367.

Joubert, F. J. and Taljaard, N. (1980). The complete primary structure of toxin CM-1b from *Hemachatus haemachatus* (Ringhals) snake venom. *Toxicon 18,* 191-198.

Joubert, F. J. and Van der Walt, S. J. (1975). *Naja melanoleuca* (Forest cobra) venom. Purification and some properties of phospholipase A. *Biochim. biophys. Acta 359,* 317.

Karlsson, E., Eaker, D., Fryklund, L., and Kadin, S. (1972). Chromatographic separation of *Enhydrina schistosa* (common sea snake) venom and the characterization of two principal neurotoxins. *Biochemistry 11,* 4628.

Kaudern, W. (1922). Sauropsides aus Madagascar, Reptiles. *Zool Jahrb. Syst. B45,* 416-458.

Kaup, J. J. (1830). *Trogonophis*. Eine neue Amphibiengattung, den Amphisbaenen zunachst verwandt. *Isis von Oken 23*, 880-881.

Khalil, F. (1951). Excretion in reptiles. IV. Nitrogenous constituents of the excreta of lizards. *J. Biol. Chem. 189*, 443-445.

Khalil, F. and Abdel-Messeih, G. (1954). Water content of tissues of some desert reptiles and mammals. *J. Exp. Zool. 125*, 407-414.

Khalil, F. and Abdel-Messeih, G. (1959a). Water, nitrogen and lipid content of the tissues of *Varanus griseus* Daudin. *Z. vergl. Physiol. 42*, 403-409.

Khalil, F. and Abdel-Messeih, G. (1959b). Effect of starvation on contents of water, nitrogen and lipids of tissues of *Varanus griseus* Daudin. *Z. vergl. Physiol. 42*, 410-414.

Khalil, F. and Abdel-Messeih, G. (1959c). The storage of extra water by various tissues of *Varanus griseus* Daudin. *Z. vergl. Physiol. 42*, 415-421.

Khalil, F. and Abdel-Messeih, G. (1961a). The storage of extra water by various tissues of *Uromastyx aegyptia* (Forskal). *Z. vergl. Physiol. 45*, 78-81.

Khalil, F. and Abdel-Messeih, G. (1961b). Effect of water deficit and water excess on the composition of blood of *Varanus griseus* Daudin. *Z. vergl. Physiol. 45*, 82-87.

Khalil, F. and Haggar G. (1955). Ureotelism and uricotelism in tortoises. *J. Exp. Zool. 130*, 423-432.

Khalil, F. and Hussein, M. F. (1962). Studies on the temperature relationships of Egyptian desert reptiles, IV. On the retention of heat of *Uromastyx aegyptia*, *Agama pallida* and *Chalcides serpoides*. *Bull. Zool. Soc. Egypt 17*, 80-88.

Khalil, F. and Hussein, M. F. (1963). Ecological studies in the Egyptian deserts. III. Daily and annual cycles of *Uromastyx aegyptia*, *Agama pallida* and *Chalcides serpoides* with special reference to temperature and relative humidity. *Proc. Zool. Soc. U. A. R. 1*, 93-108.

Khalil, F. and Yanni, M. (1959). Studies on carbohydrates in reptiles. I. Glucose in body fluids of *Uromastyx aegyptia*. *Z. vergl. Physiol. 42*, 192-198.

Khalil, F. and Yanni, M. (1961). Studies on carbohydrates in reptiles. III. Seasonal changes in glycogen content of tissues, and relative weights of organs of *Uromastyx aegyptia*. *Z. vergl. Physiol. 44*, 355-362.

Kim, Y. J., Gorman, G. C. and Huey, R. B. (1978). Genetic variation and differentiation in two species of the fossorial African skink *Typhlosaurus* (Sauria: Scincidae). *Herpetologica 34*, 192-194.

Kirk, R. L. and Hogben, L. (1945). Studies on temperature regulation, II. Amphibia and reptiles. *J. Exp. Biol. 22*, 213-220.

Klausewitz, W. (1954). Eidonomische Untersuchungen uber die Rassenkreise *Agama cyanogaster* und *Agama atricollis*. 1. Der Rassenkreis der *Agama cyanogaster*. *Senckenberg biol. 35*, 137-146.

Klausewitz, W. (1957). Eidonomische Untersuchungen uber die Rassenkreise *Agama cyanogaster* und *Agama atricollis*. 2. Die Unterarten von *Agama atricollis*. *Senchenberg biol. 38*, 157-174.

Klawe, W. L. (1964). Food of the black-and-yellow sea snake, *Pelamis platurus*, from Ecuadorian coastal water. *Copeia* 712-713.

Kluge, A. G. (1965). The systematic status of certain Australian lizards of the family Geckkonidae. *Aust. Zool. 13*, 121-125.

Knoepffler, L. Ph. (1968). Clef de determination des serpents actuellement connus du Gabon et des contrees limitrophes (Moyen-Congo, Rio Muni, Cabinda et Sud Cameroun forestier). *Biol. Gabon 4*, 183-194.

Kochva, E. (1962). On the lateral jaw musculature of the Solenoglypha with remarks on some other snakes. *J. Morph. 10*, 227-284.

Kochva, E. and Gans, C. (1965). The venom gland of *Vipera palaestinae* with comments on the glands of some other viperines. *Acta Anat. 62*, 365-401.

Kochva, E., Shayer-Wollberg, M. and Sobol, R. (1967). The special pattern of the venom gland in *Atractaspis* and its bearing on the taxonomic status of the genus. *Copeia* 763-772.

Kochva, E. and Wollberg, M. (1970). The salivary glands of Aparallactinae (Colubridae) and the venom glands of *Elaps* (Elapidae) in relation to the taxonomic status of this genus. *Zool. J. Linn. Soc. 49,* 217-224.

Kolar, K. (1955). Herpetologisches aus Sudalgerien. *Aquar. Terrar. Zeitschrift 8,* 75-78.

Komnick, H. (1970). Fine structure of the salt gland epithelium of *Uromastix acanthinurus. Verhand. Deut. Zool. Gesell. 64,* 118-122.

Kopeyan, C., Rietschoten, J. von, Martinez, G., Rochat, H., Miranda, F. and Lissitzky, S. (1973). Characterisation of five neurotoxins isolated from the venoms of two Elapidae snakes *Naja haje* and *Naja nigricollis. Eur. J. Biochem. 35,* 244-250.

Kornalik, F. (1963). Uber den Einfluss von *Echis carinatus*—toxin auf die Blutgerinnung. *Folia haematol. 80,* 73-78.

Kornalik, F., Erbanova, Z. and Mebs, D. (1978). Coagulant activity of *Thelotornis kirtlandii* venomous gland extract. *Toxicon 16,* 422 (Abst.)

Kornalik, F., Mebs, D. and Taborska, E. (1978). Coagulant, pharmacological and biochemical properties of *Thelotornis kirtlandii* venom gland extract. *Period. biol. 80,* 79-90.

Kornalik, F. and Taborska, E. (1972). Intraspecies individual variability in the composition of *Echis carinatus* venom. *Toxicon 10,* 529.

Kornalik, F. and Taborska, E. (1978). Procoagulant and defibrinating potency of the venom gland extract of *Thelotornis kirtlandii. Thromb. Res. 12,* 991-1000.

Kornalik, F., Taborska, E. and Mebs, D. (1978). Pharmacological and biochemical properties of a venom gland extract from the snake *Thelotornis kirtlandii. Toxicon 16,* 535-542.

Kramer, E. and Schnurrenberger, H. (1963). Systematik, Verbreitung und Okologie der Libyschen Schlangen. *Rev. Suisse Zool. 70,* 453-568.

Kramer, J. E. and Richardson, J. I. (1979). Volumetric reduction in nest contents of Loggerhead Sea Turtles (*Caretta caretta*) (Reptilia, Testudines, Chelonidae) on the Georgia coast. *J. Herpet. 13,* 255-260.

Kratzer, H. (1965). Uber die Tanganyika-Wasserkobra (*Boulengerina annulata stormsi*). *Salamandra 1,* 61-67.

Kritzinger, C. C. (1946). The cranial anatomy and kenesis of the South African amphisbaenid *Monopeltis capensis* Smith. *S. Afr. J. Sci. 42,* 175-204.

Kropach, C. (1971). Sea snake (*Pelamis platurus*) aggregations on slicks in Panama. *Herpetologica 27,* 131-135.

Kropach, C. (1972). *Pelamis platurus* as a potential colonizer of the Caribbean Sea. *Bull. Biol. Soc. Washington 2,* 267-269.

Kropach, C. (1975). The yellow-bellied sea snake, *Pelamis*, in the Eastern Pacific. In, Dunson, W. A. (editor), *Biology of Sea Snakes.* Univ. Park Press. pp. 185-213.

Kuhl, H. (1820). *Beitrage zur Zoologie und vergleichenden Anatomie.* Frankfurt-am-Main *1,* 1-152.

Kyegombe, D. B., Njoroge, D. K., Dossaji, S. F. and Telang, B. V. (1978). Identification and estimation of an acetylcholine-like substance in the venom of *Dendroaspis jamesoni. Acta Biol. Acad. Sci. Hungary 28,* 443-450.

Labib, R. S., Halim, H. Y. and Farag, N. W. (1979). Fractionation of *Cerastes cerastes* and *Cerastes vipera* snake venoms by gel filtration and identification of some enzymatic and biological activities. *Toxicon 17,* 337-345.

Lacepede, B. E. C. de (1789). *Histoire naturelle des Quadrupedes Ovipares et des Serpents.* Paris, *1,* 1-651.

Lakjer, T. (1926). Studien uber die trigeminus-versorgte Kaumuskulatur der Sauropsiden. *C. A. Rietzel Copenhagen.*

Lampe, E. (1911). Erster Nachtrag zum Katalog der Reptilien-und Amphibien-Sammlung des Naturhistorischen Museums der Stadt Wiesbaden. *Jahrb. Nassau Ver. Naturk. (Wiesbaden) 64,* 137-236.

Lampe, E. and Lindholm, W. A. (1901). Catalog der Reptilien-Sammlung (Schildkroten, Crocodile, Eidechsen und Chamaeleons) des Naturhistorischen Mueseums zu Wiesbaden. *Jahrb. Nassau Ver. Naturk (Wiesbaden) 54,* 177-218.

Langebartel, D. A. (1968). The hyoid and its associated muscles in snakes. *Illinois Biol. Monogr. 38,* 1-156.

Lanza, B. (1957). Su alcuni *Chalcides* del Marocco (Reptilia, Scincidae). *Monit. zool. ital. 65,* 85-98.

Lanza, B. (1964). Il genere *Sphalerosophis* e descrizione di una nuova specie (Reptilia, Serpents). *Monit. zool. ital. 72,* 47-64.

Lanza, B. (1978a). On some new or interesting East African Amphibians and Reptiles. *Monit. zool. ital. (N. S.) 10,* 229-297.

Lanza, B. (1978b). A new Somali *Hemidactylus* (Reptilia Gekkonidae). *Monit. zool. ital. (N. S.) 11,* 111-117.

Lanza, B. (1978c). *Mabuya ferrarai,* a new scincoid lizard from Somalia. *Monit. zool. ital. (N. S.) 11,* 271-280.

Lanza, B. (1979a). *Lygosoma simonettai,* a new black-headed skink from Somalia (Reptilia Scincidae). *Monit. zool. ital. (N. S.) 12,* 25-32.

Lanza, B. (1979b). *Elapsoidea chelazzii,* a new elapid snake from Somalia. *Monit. zool. ital. (N. S.) 12,* 237-245.

Lanza, B. and Carfi, S. (1966). Note su alcuni Scincidae della Somalia e descrizione di una nuova specie di *Mochlus* (Reptilia, Squamata). *Monit. zool. ital. 74,* 34-43.

Lanza, B. and Carfi, S. (1968). Gli scincidi della Somalia (Reptilia, Squamata). *Monit. zool. ital. (N. S.) 2,* 207-260.

Lanza, B. and Poggesi, M. (1975). On a new *Eremias* from central Somalia. *Monit. zool. ital. (N. S.) 6,* 305-312.

Lanza, B. and Vanni, S. (1976). On a small collection of reptiles from North Zaire. *Monit. zool. ital. (N. S.) 8,* 129-160.

Lataste, F. (1880). Diagnoses des reptiles nouveaux d'Algerie. *Naturaliste 1,* 299-325.

Lataste, F. (1881). Diagnoses des reptiles nouveaux d'Algerie. *Naturaliste 3,* 357-359.

Laurent, R. F. (1945). Contribution a la connaissance du Genre *Atractaspis* A. Smith *Revue Zool. Bot. Afr. 38,* 312-343.

Laurent, R. F. (1947a). Notes sur quelques reptiles appartenant a la collection du Musee Royale d'Histoire Naturelle de Belgique. I. Formes Africaines. *Bull. Mus. r. Hist. nat. Belgique 23,* 1-12.

Laurent, R. F. (1947b). Notes sur les Amphisbaenidae d'Afrique. *Revue Zool. Bot. Afr. 40,* 52-63.

Laurent, R. F. (1950a). Reptiles nouveaux des Kundelungu. *Revue Zool. Bot. Afr. 43,* 349-352.

Laurent, R. F. (1950b). Revision du genre *Atractaspis* A. Smith. *Inst. Roy. Sci. nat. Belgique, Mem. (ser. 2) 38,* 1-49.

Laurent, R. F. (1950g). Reptiles et Batraciens de la region de Dundo (Angola). Premiere note. *Publ. cult. Comp. Diamantes Mus. Dundo, Lisbon* 125-136.

Laurent, R. F. (1951). Deux reptiles et onze batraciens nouveau d'Afrique centrale. *Revue. Zool. Bot. Afr. 44,* 360-381.

Laurent, R. F. (1952). Reptiles et batraciens nouveaux du massif du mont Kabobo et du plateau des Marungu. *Revue Zool. Bot. Afr. 46*, 18-34.

Laurent, R. F. (1954). Reptiles et batraciens de la region de Dundo (Angola). Deuxieme note. *Publ. Cultura Cia Diamant Angola 23*, 35-84.

Laurent, R. F. (1955). Diagnoses preliminaires de quelques serpents venimeux. *Revue Zool. Bot. Afr. 51*, 127-139.

Laurent, R. F. (1956a). Contribution a l'herpetologie de la region des Grands Lacs de l'Afrique centrale. *Ann. Mus. r. Congo Belge (Sci. Zool.) 48*, 1-390.

Laurent, R. F. (1956b). Notes herpetologiques africaines. *Rev. Zool. Bot. Afr. 52*, 229-256.

Laurent, R. F. (1958). Notes herpetologiques africaines II. *Rev. Zool. Bot. Afr. 58*, 115-128.

Laurent, R. F. (1960). Notes complementaries sur les cheloniens et les ophidiens du Congo, oriental. *Ann. Mus. r. Congo. Belge (Sci. Zool.) 84*, 1-86.

Laurent, R. F. (1964a). A new subspecies of *Varanus exanthematicus* (Sauria, Varanidae). *Breviora 199*, 1-9.

Laurent, R. F. (1964b). Reptiles et amphibiens de l'Angola (Troisieme contribution). *Mus. Dundo Publ. Cult. 67*, 1-165.

Laurent, R. F. (1964c). A revision of the *punctatus* group of African *Typhlops* (Reptilia: Serpentes). *Bull. Mus. comp. Zool. 130*, 387-444.

Laurent, R. F. (1965). Contribution a l'Histoire de l'herpetologie congolaise et bibliographie generale. *Koninkl. Acad. Overz. Wetensch., Kl. Nat. Geneesk Wetensch (N. S.) 16*, 1-53.

Laurent, R. F. (1968). A re-examination of the snake genus *Lycophidion* D. & B. *Bull. Mus. comp. Zool. 136*, 461-482.

Laurenti, J. N. (1768). *Specimen medicum, exhibens synopsin reptilium emendatum cum experimentis circa venana et antidota reptilium austriacarum.* Vienna pp. 214.

Lawrence, R. F. (1959). The sand-dune fauna of the Namib desert. *S. Afr. J. Sci. 55*, 233-239.

Lee, C. Y., Chen, Y. M. and Mebs, D. (1976). Chromatographic seperation of the venom of Egyptian black snake (*Walterinnesia aegyptia*) and pharmacological characterization of its components. *Toxicon 14*, 275-281.

Lee, C. Y., Ho, C. L. and Eaker, D. (1977). Cardiotoxin-like action of a basic phospholipase A isolated from *Naja nigricollis* venom. *Toxicon 15*, 355-356.

Lee, S. R. (1976). Acetylcholinesterase isozymes from the venom of *Naja melanoleuca*. *Fed. Proc. 35*, 1656.

Leeson, F. (1950). *Identification of Snakes of the Gold Coast.* Crown Agents, London pp. 130.

Lefrou, G. and Martignoles, I. (1954). Contribution a l'etude des properties du venin d'un viperide africain: *Echis carinatus*. *Ann. Inst. Pasteur Paris 86*, 446-457.

Leloup, P. (1964). Observations sur la reproduction du *Dendroaspis jamesoni kaimosae* (Loveridge). *Bull. Soc. Roy. Zool. Anvers. 33*, 13-27.

Lemire, M., Deloince, R. and Grenot, C. (1970). Etude des cavites et glandes nasales du lezard Fouette-Queue, *Uromastix acanthinurus* Bell. *C. r. Acad. Seanc. Paris 270D*, 817-820.

Lemire, M., Deloince, R. and Grenot, C. (1972). Etude histoenzymologique de la glande "a sels" du lezard saharien *Uromastix acanthinurus* Bell. *C. r. Acad. Seanc. Paris 274D*, 3413-3416.

Leston, D. (1970). Some snakes from the forest zone of Ghana. *Brit. J. Herpet. 4*, 141-144.

Leston, D. and Hughes, B. (1968). The snakes of Tafoo, a forest cocoa-farm locality in Ghana. *Bull. I. F. A. N. A30*, 737-770.

Levinson, S. R., Evans, M. H. and Groves, F. (1976). A neurotoxic component of the venom from Blandings tree snake (*Boiga blandingi*). *Toxicon 14*, 307-312.

Leviton, A. E. and Anderson, S. C. (1963). Third contribution to the herpetology of Afghanistan. *Proc. California Acad. Sci. (ser. 4) 31*, 329-339.

Leviton, A. E. and Anderson, S. C. (1970). Review of the snakes of the genus *Lytorhynchus*. *Proc. California Acad. Sci. (ser. 4) 37*, 249-274.

Leviton, A. E. and Anderson, S. C. (1972). Description of a new species of *Tropiocolotes* (Reptilia: Gekkonidae) with a revised key to the genus. *Occ. Pap. California Acad. Sci. 96*, 1-7.

Lichtenstein, H. (1823). *Verzeichniss der Doubletten des zoologischen Museums der Konigl. Universitat zu Berlin*. Berlin pp. 118.

Lin. E. J. I. (1979). Fatbody and liver cycles in two tropical lizards *Chamaeleo hohneli* and *Chamaeleo jacksoni* (Reptilia, Lacertilia, Chamaeleonidae). *J. Herpet. 13*, 113-117.

List, J. C. (1966). Comparative osteology of the snake families Typhlopidae and Leptotyphlopidae. *Illinois Biol. Monogr. 36*, 1-112.

Lonnberg, E. (1911). Reptiles, batrachians and fishes collected by the Swedish Zoological Expedition to British East Africa, 1911. *K. Svenska Vetensk-Akad. Handl. 47*, 1-42.

Lonnberg, E. (1922). Sammlung der schwedischen Elgon-Expedition im Jahre 1920. 6. Reptiles. *Arkiv. Zool. 14*, 1-8.

Lonnberg, E. and Andersson, L. G. (1913). On a collection of reptiles from Kismayu. *Arkiv. Zool. 8*, 1-6.

Louw, A. (1974a). Snake venom toxins. The purification and properties of five non-neurotoxic polypeptides from *Naja mossambica mossambica* venom. *Biochim. biophys. Acta 336*, 470.

Louw, A. (1974b). Snake venom toxins. The amino acid sequences of three cytotoxins homologue from *Naja mossambica mossambica* venom. *Biochim. biophys. Acta 336*, 481.

Louw, A. (1974c). Snake venom toxins. Complete amino acid sequence of cytotoxin $V^{II}4$ from the venom of *Naja mossambica mossambica*. *Biochem. biophys. Res. Commun. 58*, 1022.

Louw, G. N. and Holm, E. (1972). Physiological, morphological and behavioral adaptions of the ultrapsammophilous Namib Desert Lizard *Aporosaura anchietae* (Bocage). *Madoqua (ser. 2) 1*, 67-85.

Loveridge, A. (1920). Notes on East African Lizards collected 1915-1919, with description of a new Genus and species of skink and a new subspecies of Gecko. *Proc. zool. soc. London* 131-167.

Loveridge, A. (1923a). A list of the lizards of British Territories in East Africa with keys for the diagnosis of species. *Proc. zool. soc. London* 841-863.

Loveridge, A. (1923b). Notes on East African lizards collected 1920-1923, with the description of two new races of *Agama lionotus* Blgr. *Proc. zool. soc. London* 935-969.

Loveridge, A. (1924). Checklist of the Reptilia recorded from the British Territories in East Africa. *J. E. Africa Uganda Nat. Hist. Soc. Spec. Suppl. 3*, 1-16.

Loveridge, A. (1929a). Blind snakes and pythons of East Africa. *Bull. Antiv. Inst. Amer. 3*, 14-19.

Loveridge, A. (1929b). East African reptiles and amphibians in the United States National Museum. *Natl. Mus. Nat. Hist. Bull. 151*, 1-135.

Loveridge, A. (1931a). On two amphibious snakes of the Central African lake region. *Bull. Antiv. Inst. Amer. 5*, 7-12.

Loveridge, A. (1931b). A new snake of the genus *Typhlops* from the Belgian Congo. *Copeia* 92-93.

Loveridge, A. (1932a). New reptiles and amphibians from Tanganyika Territory and Kenya Colony. *Bull. Mus. Comp. Zool. 72*, 375-387.

Loveridge, A. (1932b). New opisthoglyphous snakes of the genera *Crotaphopeltis* and *Trimerorhinus* from Angola and Kenya Colony. *Proc. Biol. Soc. Washington 45*, 83-86.

Loveridge, A. (1933). Reports on the scientific results of an expedition to the South Western Highlands of Tanganyika-Territory. VII. Herpetology. *Bull. Mus. Comp. Zool. 74*, 197-416.

Loveridge, A. (1935a). Scientific results of an expedition to rain forest regions in Eastern Africa. I. New reptiles. *Bull. Mus. Comp. Zool. 79*, 1-19.

Loveridge, A. (1935b). New geckos of the genus *Lygodactylus* from Somaliland, Sudan, Kenya and Tanganyika. *Proc. Biol. Soc. Washington 48*, 195-200.

Loveridge, A. (1936b). Revision of the Afrcan geckos of the genus *Cnemaspis*, with description of a new race. *Proc. zool. soc. London* (for 1935) 817-822.

Loveridge, A. (1936c). New geckos of the genus *Hemidactylus* from Zanzibar and Manda islands. *Proc. Biol. Soc. Washington 49*, 59-62.

Loveridge, A. (1936d). New tree snakes of the genera *Thrasops* and *Dendraspis* from Kenya Colony. *Proc. Biol. Soc. Washington 49*, 63-66.

Loveridge, A. (1936e). African reptiles and amphibians in the Field Museum of Natural History. *Field Mus. Nat. Hist. (ser. zool.) 22*, 1-111.

Loveridge, A. (1936f). Scientific results of an expedition to rain forest regions in eastern Africa. V. Reptiles. *Bull. Mus. Comp. Zool. 79*, 209-337.

Loveridge, A. (1937). Zoological results of the George Vanderbildt African expedition of 1934. VII. Reptiles and amphibians. *Proc. Acad. nat. Sci. Philadelphia 89*, 265-296.

Loveridge, A. (1939). Revision of the African Snakes of the genera *Mehelya* and *Gonionotophis*. *Bull. Mus. Comp. Zool. 86*, 131-162.

Loveridge, A. (1940). Revision of the African snakes of the genera *Dromophis* and *Psammophis*. *Bull. Mus. Comp. Zool. 87*, 1-70.

Loveridge, A. (1941a). Cerain Afro-America geckos of the genus *Hemidactylus*. *Copeia* 245-248.

Loveridge, A. (1941b). New geckos (*Phelsuma* and *Lygodactylus*), snake (*Leptotyphlops*) and frog (*Phrynobatrachus*) from Pemba Island, East Africa. *Proc. Biol. Soc. Washington 54*, 175-178.

Loveridge, A. (1941c). Revision of the African lizards of the family Amphisbaenidae. *Bull. Mus. Comp. Zool. 87*, 353-451.

Loveridge. A. (1941d). Report on the Smithsonian-Firestone Expedition's collection of reptiles and amphibians from Liberia. *Proc. U. S. Natl. Mus. 91*, 113-140.

Loveridge, A. (1941e). Revision of the African terrapins of the Family Pelomedusidae. *Bull. Mus. Comp. Zool. 88*, 467-524.

Loveridge, A. (1942a). Revision of the Afro-Oriental Geckos of the Genus *Phelsuma*. *Bull. Mus. Comp. Zool. 89*, 437-482.

Loveridge, A. (1942b). Revision of the African lizards of the family Gerrhosauridae. *Bull. Mus. Comp. Zool. 89*, 483-543.

Loveridge, A. (1942c). Scientific results of a fourth expedition to forested areas in east and central Africa. IV. Reptiles. *Bull. Mus. Comp. Zool. 91*, 237-373.

Loveridge, A. (1944a). New geckos of the genera *Afroedura* new genus and *Pachydactylus* from Angola. *Amer. Mus. Novitate 1254*, 1-4.

Loveridge, A. (1944b). Revision of the African lizards of the family Cordylidae. *Bull. Mus. Comp. Zool. 95*, 1-118.

Loveridge, A. (1944c). Further revisions of African snake genera. *Bull. Mus. Comp. Zool. 95*, 121-247.

Loveridge, A. (1944d). Remarks on the gekkonid genera *Homopholis* and *Platypholis* with description of a new race. *Proc. Biol. Soc. Washington 57*, 1-4.

Loveridge, A. (1946). A new worm-lizard (*Ancylocranium barkeri*) from Tanganyika Territory. *Proc. Biol. Soc. Washington 59*, 73-74.

Loveridge, A. (1947). Revision of the African lizards of the family Gekkonidae. *Bull. Mus. Comp. Zool. 98*, 1-469.

Loveridge, A. (1948). The lizard *Amphisbaena kraussi* rediscovered in the Gold Coast. *Copeia* 213-214.

Loveridge, A. (1951a). On reptiles and amphibians from Tanganyika Territory collected by C. J. P. Ionides. *Bull. Mus. Comp. Zool. 106*, 177-204.

Loveridge, A. (1951b). Synopsis of the African green snakes *Philothamnus* and *Chlorophis* with the description of a new form. *Bull. Inst. Roy. Sci. nat. Belgique 27*, 1-12.

Loveridge, A. (1952a). Mission A. Villiers au Togo et au Dahomey (1950). XII. Tortoises and lizards. *Bull. I. F. A. N. 14*, 229-242.

Loveridge, A. (1953). Zoological results of a fifth expedition to East Africa. III. Reptiles from Nyasaland and Tete. *Bull. Mus. Comp. Zool. 110*, 141-322.

Loveridge, A. (1955). On a second collection of reptiles and amphibians taken in Tanganyika Territory by C. J. P. Ionides, Esq. *J. E. Afr. Nat. Hist. Soc. 22*, 169-198.

Loveridge, A. (1956a). A new subgenus of *Chamaeleo* from Rhodesia and new race of *Mabuya* from Keya Colony. *Breviora 59*, 1-4.

Loveridge, A. (1956b). On a third collection of reptiles taken in Tanganyika by C. J. P. Ionides Esq. *Tanganyika Notes and Records 43*, 1-19.

Loveridge, A. (1957). Check-list of the reptiles and amphibians of East Africa (Uganda, Kenya, Tanganyika, Zanzibar). *Bull. Mus. Comp. Zool. 117*, 153-362.

Loveridge, A. (1958). Revision of five African snake genera. *Bull. Mus. Comp. Zool. 119*, 1-198.

Loveridge, A. (1962). New worm-lizards (*Ancylocranium* and *Amphisbaena*) from southeastern Tanganyika Territory. *Breviora 163*, 1-6.

Loveridge, A. and Williams, E. E. (1957). Revision of the African tortoises and turtles of the suborder Cryptodira. *Bull. Mus. Comp. Zool. 115*, 161-557.

Mackay, N., Ferguson, J. C., Ashe, J., Bagshawe, A., Forrester, A. T. T. and McNicol, G. P. (1969). The venom of the boomslang (*Dispholidus typus*): *in vivo* and *in vitro* studies. *Thromb. Diath. haemmorrh. 21*, 234.

Mackay, N., Ferguson, J. C. and McNicol, G. P. (1970). Effects of the venom of the rhinoceros horned viper (*Bitis nasicornis*) on blood coagulation, platelet aggregation and fibrinolysis. *J. Clinical Path. 23*, 789-796.

Madsen, T., Lundstrom, H. and Fohlman, J. (1979). Purification of monospecific antisera against the venom of the Cape cobra (*Naja nivea*). *Toxicon 17*, 326-330.

Mahendra, B. C. (1935a). On the peculiar apertures in the vertebral centra of *Hemidactylus flaviviridis* Ruppel. *Current Sci. Bangalore 4*, 34.

Mahendra, B. C. (1935b). Sexual dimorphism in the Indian House-gecko, *Hemidactylus flaviviridis* Ruppel. *Current Sci. Bangalore 4*, 178-179.

Mahendra, B. C. (1936). Contributions to the Bionomics, Anatomy, Reproduction and Development of the Indian House-gecko, *Hemidactylus flaviviridis* Ruppel. Part I. *Proc. Indian Acad. Sci. 4*, 250-281.

Mahendra, B. C. (1941). Contributions to the Bionomics, Anatomy, Reproduction and Development of the Indian House-gecko. *Hemidactylus flaviviridis* Ruppel. part II. The problem of locomotion. *Proc. Indian Acad. Sci. 13*, 288-306.

Mahendra, B. C. (1942). Contributions to the Bionomics, Anatomy, Reproduction and Development of the Indian House-gecko, *Hemidactylus flaviviridis* Ruppel. part III. The Heart and Venous System. *Proc. Indian Acad. Sci. 15*, 231-252.

Manacas, S. (1951). Saurios da Guine Portuguesa. *An. Junta Invest. Coloniais 6*, 53-67.

Manacas, S. (1952). Saurios de Mocambique. *An. Junta Invest. Ultramar (zool.) 7*, 130-159.

Manacas, S. (1955). Saurios e ofidios da Guine Portuguesa. *An. Junta Invest. Ultramar (zool.) 10*, 1-29.

Manacas, S. (1957). Contribuicao para o estudo dos repteis (saurios e leptotiflopideos) de Mocambique. *An Junta Invest. Ultramar (zool.) 12*, 1-8.

Marsh, N. A. and Glatston, A. (1974). Some observations on the venom of the Rhinoceros horned viper, *Bitis nasicornis* (Shaw). *Toxicon 12*, 621.

Marsh, N. A. and Whaler, B. C. (1974). Seperation and partial characterization of a coagulant enzyme from *Bitis gabonica* venom. *Brit. J. Haematology 26*, 295-306.

Marshall, A. J. and Hook, R. (1960). The breeding biology of equatorial vertebrates: reproduction of the lizard *Agama agama lionotus* Boulenger at Lat. 0°01'N. *Proc. zool. soc. London 134*, 197-205.

Martin-Moutot, N. and Rochat, H. (1979). Isolation and characterization of a toxic phospholipase A2 in the spitting cobra (*Naja mossambica mossambica*) venom. *Toxicon 17*, 127-136.

Marx, H. (1956). A new lacertid lizard from Angola. *Fieldiana Zool. 39*, 5-9.

Marx, H. (1958a). Sexual dimorphism in coloration in the viper *Cerastes vipera* L. *Nat. Hist. Miscell. 164*, 1-2.

Marx. H. (1958b). Egyptian snakes of the genus *Psammophis*. *Fieldiana Zool. 39*, 191-200.

Marx, H. (1959). Review of the Colubrid snake genus *Spalerosophis*. *Fieldiana Zool. 39*, 347-361.

Marx, H. (1968). Checklist of the reptiles and amphibians of Egypt. *Spec. Publ. U. S. Naval Med. Res. Unit 3, Cairo* 1-91.

Marx, H. and Rabb, G. B. (1965). Relationships and Zoogeography of the Viperine Snakes (Family Viperidae). *Fieldiana Zool. 44*, 161-206.

Matschie, P. (1893). Die Reptilien und Amphibien Togogebietes. *Mitt. Forsch. Gel. Dt. Schutzgeb. 6*, 207-215.

Matz, G. (1973). *Mabuya brevicollis* (Wiegmann). *Aquarama 7*, 54.

Matz, G. (1974). Les boides ou serpents constricteurs. 4. *Eryx* Daudin 1803. *Aquarama 8*, 53-55.

McDowell, S. B. (1964). Partition of the genus *Clemmys* and related problems in the taxonomy of the aquatic Testudinidae. *Proc. zool. soc. London 143*, 239-279.

McDowell, S. B. (1967). Osteology of the Typhlopidae and Leptotyphlopidae: a critical review. *Copeia* 686-692.

McDowell, S. B. (1968). Affinities of the snakes usually called *Elaps lacteus* and *E. dorsalis. J. Linn. Soc. (zool.) 47*, 561-578.

McDowell, S. B. (1974). A catalogue of the snakes of New Guinea and the Solomons, with special reference to those in the Bernice P. Bishop Museum. Part I. Scolecophidia. *J. Herpet. 8*, 1-57.

McDowell, S. B. (1975). A catalogue of the snakes of New Guinea and the Somons, with special reference to those in the Bernice P. Bishop Museum. Part II. Anilioidea and Pythoninae. *J. Herpet. 9*, 1-80.

McDowell, S. B. (1979). A catalogue of the snakes of New Guinea and the Solomons, with special reference to those in the Bernice P. Bishop Museum. Part III. Boinae and Acrochordoidea (Reptilia, Serpentes). *J. Herpet. 13*, 1-92.

McLachlan, G. R. and Spence, J. M. (1966a). Notes on *Cordylus macropholis* Boulenger. *Ann. Cape Prov. Mus. 5*, 139-143.

McLachlan, G. R. and Spence, J. M. (1966b). The genus *Pachydactylus* (Part 1). *Ann. Cape Prov. Mus. 5*, 149-156.

McLachlan, G. R. and Spence, J. M. (1967). A new species of *Pachydactylus* from Sesfontein, South West Africa. *Cimbebasia 21*, 3-8.

Mebs, D. (1969). Uber Schlangengift-Kallikreine: heinigung und Eigenschaften eines Kininfreisetzenden Enzyms ans dem Gift der Viper *Bitis gabonica. Hoppe-Seyler's Z. Physiol. Chem. 350*, 1563-1569.

Mebs, D. (1970). Biochemistry of kinin-releasing enzymes in the venom of the viper *Bitis gabonica* and of the lizard *Heloderma suspectum. Adv. expl. Biol. 8*, 107.

Meek, S. E. (1910). Batrachians and reptiles from British East Africa. *Field Mus. Nat. Hist., zool. ser. 7*, 403-414.

Meek, S. E. and Elliot, D. G. (1897). List of Fishes and Reptiles obtained by Field Columbian Museum East African Expedition to Somaliland in 1896. *Field Mus. Nat. Hist., zool. ser. 1*, 161-184.

Meier, H. (1977). Beobachtungen an *Phelsuma standing* (Reptilia, Sauria, Gekkonidae). *Salamandra 13*, 1-12.

Meij, H. S. and Meyer, B. J. (1977). Effects of cobra (*Naja nivea*) neurotoxins on presynaptic inhibition in spinal cord of frogs. *S. Afr. J. Sci. 73*, 119-121.

Menetries, E. (1832). *Catalogue raisonnne des objects de Zool. recneillis dans un voyage an Caucase et jusqu'aux frontieres actuelles de la Perse.* St. Petersburg.

Menzies, J. I. (1958). Breeding behaviour of the chamaeleon (*Chamaeleo gracilis*) in Sierra Leone. *Brit. J. Herpet. 2*, 130-132.

Menzies, J. I. (1966). The snakes of Sierra Leone. *Copeia* 169-179.

Merkle, D. A. (1975). A taxonomic analysis of the *Clemmys* complex (Reptilia, Testudines) utilizing starch gel electrophoresis. *Herpetologica 31*, 162-166.

Merrem, B. (1820). *Versuch eines Systems der Amphibien. Testamen systematis Amphibiorum.* Marburg pp. 189.

Merten, R. (1931). *Ablepharus boutonii* (Desjardin) und seine geographische variation. *Zool. Jahrb. Jena B61*, 63-193.

Mertens, R. (1934). Die Inseln-Reptilien, ihre Ausbreitung Variation und Artbildung. *Zoologica* (Stuttgart) *32*, 1-209.

Mertens, R. (1936). Eine neue Natter der Gattung *Helicops* aus Inner-Afrika. *Zool. Anz. 114*, 284-285.

Mertens, R. (1937a). Uber eine herpetologische Sammlung aus dem Gebiete des Njarasa Grabens Ost-Afrika. *Veroff. dt. Kolon.-u. Ubersee Mus. Bremen 2*, 1-9.

Mertens, R. (1937b). Zur Kenntnis der Reptilien Fauna der Luderitz-Bucht, Sudwest-Afrika. *Veroff. dt. Kolon.-u. Ubersee Mus. Bremen 2*, 10-16.

Mertens, R. (1937c). Eine neue. Tiergeographisch bemerkenswerte Eideschse aus Kamerun. *Senckenbergiana 19*, 381-385.

Mertens, R. (1937d). Reptilien und Amphibien aus dem sudlichen Inner-Afrika. *Abh. Senckenb. naturf. Ges. 435*, 1-23.

Mertens, R. (1938a). Herpetologische Ergebnisse einer Reise nach Kamerun. *Abh. Senckenb. naturf. Ges. 442*, 1-52.

Mertens, R. (1938b). Amphibien und reptilien aus Angola gesammelt von W. Schack. *Senckenbergiana 20*, 425-443.

Mertens, R. (1940). Zur herpetologie Kameruns und Deutsch-Ostafrikas. *Zool. Anz. 131*, 239-250.

Mertens, R. (1951). *Brookesia stumpffi* ein madagassischen Zwergehamaleon in Gefangenschaft. *Aquar. Terrar. Zeits. 4*, 329-330.

Mertens, R. (1953). Beobachtungen am Madagassischen Taggecko. *Phelsuma m. madagascariensis. Aquar. Terrar. Zeits. 6*, 152-155.

Mertens, R. (1954a). Neue Schlangenrassen aus Sudwest- und Sudafrika. *Zool. Anz. 152*, 213-219.

Mertens, R. (1954b). Studien uber die Reptilienfauna Madagaskars. II. Eine neue Rasse von *Phelsuma madagascariensis. Senckenberg biol. 35*, 13-16.

Mertens. R. (1955). Die Amphibien und Reptilien Sudwestafrikas. *Abh. Senckenb. naturf. Ges. 490*, 1-172.

Mertens, R. (1962a). Studien uber die Reptilienfauna Madagaskars. III: Die Arten und Unterarten der Geckonengattung *Phelsuma. Senckenberg biol. 43*, 81-127.

Mertens, R. (1962b). Die bisher lebend singefuhrten Taggeckos der Gattung *Phelsuma. Aquar. Terrar. Zeits. 15*, 148-153.

Mertens, R. (1962c). Bemerkungen uber *Uromastix acanthinurus* als Rassenkreiss (Rept. Sauria). *Senckenberg. biol. 43*, 425-432.

Mertens, R. (1963a). The geckos of the genus *Phelsuma* on Mauritius and adjacent islands. *Mauritius Inst. Bull. 5*, 299-304.

Mertens, R. (1963b). Studien uber die Reptilienfauna Madagaskars. IV: Zwei neue Arten der Geckonengattung *Phelsuma. Senckenberg. biol. 44*, 349-356.

Mertens, R. (1963c). Liste der rezenten Amphibien und Reptilien. Helodermatidae, Varanidae, Lanthanotidae. *Des. Tierreich 79*, 1-26.

Mertens, R. (1964a). Studien uber die Reptilienfauna Madagaskars. V: Funf neue Rassen der Geckonengattung *Phelsuma. Senckenberg. biol. 45*, 99-112.

Mertens, R. (1964b). Das Chamaleon der Insel Pemba. *Senckenberg. biol. 45*, 113-116.

Mertens, R. (1964c). Der Eidechsenschwanz als Haftorgan. *Senckenberg. biol. 45*, 117-122.

Mertens, R. (1966a). Liste der rezenten Amphibien und Reptilien. Chamaeleonidae. *Das Tierreich 83*, 1-37.

Mertens, R. (1966b). Die nichtmadagassischen Arten und Unterarten der Geckonengattung *Phelsuma. Senckenberg, biol. 47*, 85-110.

Mertens, R. (1968). Remarks on the scientific name of the African puff adder. *Copeia* 621-622.

Mertens, R. and Muller, L. (1928). Liste der amphibien und reptilien Europas. *Abh. Senckenb. naturf. Ges. 41*, 1-62.

Methuen, P. A. (1919). Description of a new snake from the Transvaal, together with a new diagnosis and key to the genus *Xenocalamus* and some batrachia from Madagascar. *Proc. zool. soc. London* 349-355.

Methuen, P. A. and Hewitt, J. (1913). On a collection from Madagascar during the year 1911. *Ann. mede. Transvaal. Mus. 5*, 183-186.

Methuen, P. A. and Hewitt, J. (1914). Records and descriptions of the reptiles and batrachians of the collection of Percy Sladen Memorial Expedition to Great Namaqualand 1912-1913. *Ann. Transvaal Mus. 4*, 118-145.

Methuen, P. A. and Hewitt, J. (1915). A contribution to our knowledge of the anatomy of chamaeleons. *Trans. Roy. Soc. S. Afr. 4*, 89-104.

Meyen, F. J. F. (1835). Beitrage zur Zoologie gesammelt auf einer Reise um die Erde. Siebente Abhandlung Amphibien. *Nova Acta Leopoldina 17*, 183-268.

Millot, J. (1951). Un lezard d'eau a Madagascar (*Scelotes astrolabi* Dum. et Bib.). *Le Naturaliste malagache 3*, 87-90.

Milne-Edwards, M. H. (1829). Recherches zoologiques pour servir a l'histoire de lezards extraites d'une monographie de ce genre. *Ann. Sci. Nat. 16*, 50-89.

Minton, S. A. (1968). Antigenic relationships of the venom of *Atractaspsi microlepidota* to that of other snakes. *Toxicon 6*, 59-64.

Minton, S. A., Anderson, S. C. and Anderson, J. A. (1970). Remarks on some geckos from southwest Asia, with descriptions of three new forms and a key to the genus *Tropiocolotes. Proc. California Acad. Sci. (ser. 4) 7*, 333-362.

Minton, S. A. and Salanitro, S. K. (1972). Serological relationships among some colubrid snakes. *Copeia* 246-252.

Miranda, R., Kupeyan, C., Rochat, H., Rochat, C. and Lissitzky, S. (1970). Purification of animal neurotoxins. Isolation and characterization of four neurotoxins from two different sources of *Naja haje* venom. *Euro. J. Biochem. 17*, 477-484.

Mitchell, A. J. L. and Steyn, W. (1965). *Gerrhosaurus flavigularis falvigularis* Wiegmann in South West Africa (Reptilia, Codrylidae). *Cimbebasia 12*, 13-15.

Mitchell, A. J. L. and Steyn, W. (1967). Further distribution records of reptiles in South West Africa. *Cimbebasia 21*, 23-26.

Mohamed, A. H., El-Serougi, M. and Khaled, L. Z. (1969). Effects of *Cerastes cerastes* venom on blood coagulation mechanisms. *Toxicon 7*, 181.

Mahamed, A. H. and Khaled, L. Z. (1966). Effect of the venom of *Cerastes cerastes* on nerve tissue and skeletal muscle. *Toxicon 3*, 223.

Mohamed, A. H. and Nawar, N. N. Y. (1975). Dysmelia in mice after maternal *Naja nigricollis* envenomation: a case report. *Toxicon 13*, 475-477.

Mohamed, A. H., Saleh, A. M., Ahmed, S. and Beshir, S. R. (1975). Histopathological and histochemical effects of *Naja haje* venom on kidney tissue of mice. *Toxicon 13*, 409-413.

Montgomery, J. (1959). Two cases of ophthalmoplegia due to berg adder bite. *Cent. Afr. J. Med. 5*, 173-177.

Moreau de Jonnes, A. (1818). Monographie du *mabouia* des murailles ou *gecko mabouia* des Antilles. *Bull. Soc. Philom. Paris,* 138-139.

Morita, T., Iwanaga, S. and Suzuki, T. (1976). Activation of bovine prothrombin by an activator isolated from *Echis carinatus* venom. *Thromb. Res. 8*, 59-65.

Mudrack, W. (1977). Wenn's kuhl und feucht wird, taut er auf. *Cophoscincopus durus*, ein kleiner Wasserskink. *Aquar. Mag.* (Stuttgart) *11*, 282-283.

Muller, L. (1907). Uber einen neuen Gecko aus Kamerun und eine neue Colubrine Schlage aus Centralohina. *Zool. Anz. 31*, 824-830.

Muller, L. (1909). Vorlaufige Mitteilung uber ein neues Chamaleon und ein neuen Gecko aus Kamerun. *Jahrb. Nassau Ver. Nat. (Wiesbaden) 62*, 111-115.

Muller, L. (1910). Beitrage zur Herpetologie Kameruns. *Abhand. Konig. Bayer Akad. Wiss. (ser. 2) 24*, 543-626.

Muller, L. (1922). Uber eine neue Uromastix-Art aus Zentral-Sahara. *Naturwiss. Beobachter* (Frankfurt) *63*, 193-202.

Muller, L. (1938). Uber die von den Herren W. Uthmoller und L. Bohmann im britischen Madatsgebiet "Tanganyika Territory" gesammelten Chamaleons. *Zool. Anz. 122*, 20-23.

Neumann, O. (1905). Uber nordost-afrikanische und arabische Kriechtiere. *Zool. Jahrb. Syst. 22*, 389-404.

Nickel, E. (1901). Die Kriechtiere Deutsch-Ostafrikas mit besonderer Berucksichtigung der im Museum in Frankfurt a O. befindlichen Arten. *Helios 18*, 65-73.

Nieden, F. (1910a). Neue Reptilien und Amphibien aus Kamerun. *Arch. Natur., Berlin 76*, 234-246.

Nieden, F. (1910b). *Die Fauna der deutschen Kolonien. Reihe I. Kamerun Heft. 2: Die Reptilien (auser den Schlangen) und Amphibien.* Berlin pp. 75.

Nieden, F. (1913a). Neues Verzeichnis der Kriechtiere (ausser den Schlangen) von Deutsch-Ostafrika. I. Reptilia. *Mitt. zool. Mus. Berlin 7*, 51-100.

Nieden, F. (1913b). *Chamaeleon fischeri* Rchw. und eine Unterarten. *Sitz. Ges. naturf. Freunde Berlin,* 231-249.

Noble, E. M. (1966). Occurrence of *Psammophis angolensis* in Ethiopia. *Copeia* 125-126.

Odhner, T. (1908). Reptilien und Batrachier, gesammelt von Dr. I. Tragardh in Natal und Zululand 1904-1905. *Ark. f. zool. 4*, 1-7.

Oelofsen, B. and Vorster, W. (1976). A new record for *Ptenopus* (Reptilia: Gekkonidae) from Calitzdorp. *Zool. Africana 11*, 1-22.

Orton, G. L. and Morrison, F. D. (1946). Some amphibians and reptiles from the Gold Coast. *Copeia* 15-17.

O'Shaughnessy, A. W. E. (1873). Reptilia. *Zool. Rec. 10*, 75-94.

O'Shaughnessy, A. W. E. (1875). Descriptions of new species of Gekkotidae in the British Museum Collection. *Ann. Mag. Nat. Hist. (ser. 4) 16,* 262-266.

Otis, V. (1973). Haemocytological and serum chemistry parameters of the African puff adder, *Bitis arietans. Herpetologica 29,* 110-116.

Parker, H. W. (1925). New and rare reptiles and batrachians from Madagascar. *Ann. Mag. Nat. Hist. (ser. 9) 16,* 390-394.

Parker, H. W. (1927). Parallel evolution in some Opisthoglyphous snakes, with the description of a new snake. *Ann. Mag. Nat. Hist. (ser. 9) 20,* 81-86.

Parker, H. W. (1930). Three new reptiles from Somaliland. *Ann. Mag. Nat. Hist. (ser. 10) 6,* 603-606.

Parker, H. W. (1932a). Scientific results of the Cambridge expedition to the east African lakes, 1930-31. 5. Reptiles and amphibians. *J. Lin. Soc. (Zool.) 38,* 213-229.

Parker, H. W. (1932b). Two collections of reptiles and amphibians from British Somaliland. *Proc. Zool. Soc. London* 335-367.

Parker, H. W. (1935a). A new species of amphisbaenid lizard from Bechuanaland. *Ann. Mag. Nat. Hist. (ser. 10) 15,* 582-583.

Parker, H. W. (1935b). Two new lizards from Somaliland. *Ann. Mag. Nat. Hist. (ser. 10) 16,* 525-529.

Parker, H. W. (1935c). A new melanic lizard from Transjordania, and some speculations concerning melanism. *Proc. Zool. Soc. London* 137-142.

Parker, H. W. (1936a). Dr. Karl Jordan's Expedition to South West Africa and Angola: herpetological collections. *Novit. Zool. 40,* 115-146.

Parker, H. W. (1936b). Reptiles and amphibians collected by the Lake Rudolf Rift Valley Expedition. *Ann. Mag. Nat. Hist. (ser. 10) 18,* 594-609.

Parker, H. W. (1940). Undescribed anatomical structures and new species of reptiles and amphibians. *Ann. Mag. Nat. Hist. (ser. 11) 5,* 257-274.

Parker, H. W. (1942). The lizards of British Somaliland. *Bull. Mus. Comp. Zool. 91,* 1-101.

Parker, H. W. (1949). The snakes of Somaliland and the Sokotra Islands. *Zool. Verh. Leiden 6,* 1-115.

Parsons, T. S. and Cameron, J. E. (1977). Internal relief of the digestive tract. in, *Biology of the Reptilia* (Editors, Gans, C. and Parsons, T. S.) 6, 159-223.

Pasteur, G. (1959). Note preliminaire sur les tarentes de l'Ouest africain (Sauriens, Gekkonides). *Soc. Sci. Nat. Phys. Maroc 2,* 41-42.

Pasteur, G. (1960a). Les tarentes de l'Ouest Africain. I. *Tarentola hoggarensis. Bull. Soc. Sci. Nat. Phys. Maroc 40,* 77-84.

Pasteur, G. (1960b). Notes preliminaires sur les Lygodactyles (Gekkonides). I. Remarques sur les sous-especes de *Lygodactylus picturatus. Bull. I. F. A. N. (ser. A) 22,* 1441-1452.

Pasteur, G. (1960c). Redecouverte et validite probable du Gekkonide *Tropicolotes nattereri* Steindachner. *Bull. Soc. Sci. Nat. Phys. Maroc. 40,* 143-145.

Pasteur, G. (1961). Premieres observations sur les Sauriens rapportes du Tsiafajavona par le professeur Millot. *Mem. Inst. sci. Madagascar A13,* 149-172.

Pasteur, G. (1962a). La super-espece *Chalcides mionecton-Chalcides atlantis. Bull. Soc. Sci. nat. Phys. Maroc 42,* 57-60.

Pasteur, G. (1962c). Notes preliminaires sur les Lygodactyles (Gekkonides). III. Diagnose de *Millotisaurus* gen. nov. de Madagascar. *Soc. Sci. nat. phys. Maroc 3,* 65-66.

Pasteur, G. (1964a). Notes preliminaires sur les Lygodactyles (Gekkonides) IV. Diagnose de quelques formes africaines et malgaches. *Bull. Mus. nat. Hist. nat. Paris (ser. 2) 36,* 311-314.

Pasteur, G. (1964b). Recherches sur l'evolution des Lygodactyles. Lezards afro-malgaches actuels. *Trav. Inst. scient. cherif. (ser. zool.) 29,* 1-160.

Pasteur, G. (1967a). Redecouverte d'un genre de Sauriens malgaches: *Microscalabotes* (Gekkonides). *Ann. Fac. Sci. Univ. Madagascar 5,* 75-77.

Pasteur, G. (1967b). Note preliminaire sur les Geckos du genre *Lygodactylus* rapportes par Ch. Blanc du Mont Bity (Madagascar). *Bull. Mus. nat. Hist. nat. Paris (ser. 2) 39,* 439-443.

Pasteur, G. (1967c). Un serpent endemique du Maghreb: *Sphalerosophis dolichospilus* (Werner), Colubridae. *Bull. Mus. nat. Hist. nat. Paris (ser. 2) 39,* 444-451.

Pasteur, G. (1981). A survey of the species groups of the Old World Scincid genus *Chalcides. J. Herpet. 15,* 1-16.

Pasteur, G. and Blanc, C. P. (1967). Les lezards du sous-genre malgache de Lygo-dactyles *Domerguella* (Gekkonides). *Bull. Soc. Zool. France 92,* 583-597.

Pasteur, G. and Bons, J. (1959). Les Batraciens du Maroc. *Trav. Inst. Sci. cherif. (ser. zool.) 17,* 1-241.

Pasteur, G. and Bons, J. (1960). Catalogue des reptiles actuels de Maroc. Revision de formes d'Afrique, d'Europe et d'Asia. *Trav. Inst. Sci. cherif. (ser. zool.) 21,* 1-132.

Pasteur, G. and Girot, B. (1960). Les tarentes de l'Ouest Africain II. *Tarentola mauritanica. Bull. Soc. Sci. nat. phys. Maroc 40,* 309-322.

Pasteur, G. and Paulian, R. (1962). Diagnose d'un lezard apode de Madagascar: *Pygomeles petteri* n. sp. (Scincidae). *Bull. Mus. nat. Hist. nat. Paris (ser. 2) 34,* 66.

Patel, R. and Excell, B. J. (1974). The modes of action of whole *Dendroaspis jamesoni* venom on skeletal nerve-muscle preparations. *Toxicon 12,* 577-585.

Patel, R. and Excell, B. J. (1975). The effects of lethal components of *Dendroaspis jamesoni* snake venom on neuromuscular transmission and on membrane perme-ability. *Toxicon 13,* 295-304.

Patterson, R. W. (1974). Hatching the African python, *Python sebae,* in captivity. *Inter. Zoo Yrbk. 14,* 81-82.

Patterson, R. W. (1978). Hatching of Anchieta's dwarf python, *Python anchietae. Inter. Zoo Yrbk. 18,* 99-101.

Pellegrin, J. (1909a). Reptiles et batraciens recoltes par M. Ch. Alluaud en Egypte et au Soudan egyptien. *Bull. Soc. Zool. France 34,* 203-205.

Pellegrin, J. (1909b). Reptiles du Soudan recoltes par la Mission Tilho-Gaillard. Description d'une espece nouvelle. *Bull. Mus. nat. Hist. nat. Paris 15,* 413-415.

Pellegrin, J. (1910). Mission en Mauritanie occidentale III. Partie zoologique, Reptiles. *Actes Soc. Linn. Bordeaux 44,* 21-25.

Pellegrin, J. (1912a). Reptiles, Batraciens et Poissons du Maroc (Mission de Mme. Camille du Gast). *Bull. Soc. Zool. France 37,* 255-262.

Pellegrin, J. (1912b). Reptiles, Batraciens et Poissons du Maroc (Recoltes par le Dr. H. Millet). *Bull. Soc. Zool. France 37,* 262-264.

Pellegrin, J. (1925b). Les reptiles et batraciens du Grand et du Moyen Atlas. *Comptes Rendus Acad. Sci. Paris 181,* 880-882.

Pellegrin, J. (1925c). Etude zoologiques. Batraciens et Reptiles. In, Kilian, C. *Au Hogger Mission de 1922.* Paris, pp. 141-142.

Pellegrin, J. (1926a). Liste des Reptiles, Batraciens et Poissons d'Eau douce des Collections du Musee de l'Institut scientifique cherifien a Rabat. *Bull. Soc. Sci. Nat. Maroc 5,* 315-321.

Pellegrin, J. (1926e). Mission J. Pellegrin au Maroc. Reptiles, Batraciens et Poissons. *Bull. Mus. nat. Hist. nat. paris 32,* 120-124.

Pellegrin, J. (1926f). Reptiles, Batraciens et Poissons du Maroc oriental recueillis par M. P. Pallary. *Bull. Mus. nat. Hist. nat. Paris 32,* 159-162.

Pellegrin, J. (1927a). Les reptiles et les batraciens de l'Afrique du Nord francaise. *Compte Rendu Assoc. Franc. Avanc. Sci. Constantine 51*, 260-264.

Pellegrin, J. (1928). Reptiles et Poissons du Moyen-Atlas recueillis par M. P. Pallary. *Bull. Mus. nat. Hist. nat. Paris 34*, 243-245.

Pellegrin, J. (1931). Reptiles, Batraciens et Poissons du Sahara central recueillis par le Pr. Seurat. *Bull. Mus. nat. Hist. Nat. Paris (ser. 2) 3*, 216-218.

Pellegrin, J. (1934). Reptiles, Batraciens et Poissons du Sahara central. *Mem. Soc. Hist. Nat. Afr. Nord, Alger 4*, 50-57.

Pellegrin, J. (1936). Mission au Tibesti. Etude preliminaire de la Faune du Tibesti. Reptiles, Batraciens. *Mem. Acad. Sci. Inst. France 62*, 50-52.

Perret, J. L. (1959). Etudes herpetologiques africaines. *Bull. Soc. Neuchatel Sci. Nat. 82*, 247-253.

Perret, J. L. (1960). Une nouvelle et remarquable especes d'*Atractaspis* (Viperidae) et quelques autres serpents d'Afrique. *Revue suisse Zool. 67*, 129-139.

Perret, J. L. (1975). La differenciation dans le genre *Panaspis* Cope (Reptilia, Scincidae) *Bull. Soc. Neuchatel Sci. Nat. 98*, 5-16.

Peters, J. A. and Broadley, D. G. (1967). The scientific name of the African Puff adder. *Copeia* 864-865.

Peters, U. (1969). Some observations on the captive breeding of the Madagascan tortoise *Testudo radiata* at Sydney Zoo. *Inter. Zoo Yrbk. 9*, 29.

Petit, G. (1928). Sur le *Chalarodon madagascariensis* Peters. *Bull. Soc. Zool. France 53*, 401-405.

Petter-Rousseaux, A. (1953). Recherches sur la croissance et le cycle d'activite testiculaire de *Natrix natrix helvetica* (Lacepede). *Terre et vie 4*, 175-223.

Pfeffer, G. (1893). Ostafrikanische Reptilien und Amphibien, gesammelt von Herrn Dr. F. Stuhlmann in Jahre 1888 und 1889, *Jahrb. Hamburg Wiss. Anst. 10*, 71-105.

Phillips, E. J. (1977). Raising hatchlings of the leatherback turtle. *Dermochelys coriacea. Brit. J. Herpet. 5*, 677-678.

Phisalix, M. (1914). Anatomie comparee de la tete et de l'appareil venimeux chez les serpents. *Ann. Sci. Nat. Zool. 19*, 1-114.

Phisalix, M. (1922). *Animaux venimeux et Venins*. Paris, 2 vols.

Pickwell, G. V. (1971). Knotting and coiling behavior in the pelagic sea snake *Pelamis platurus* (L.). *Copeia* 348-350.

Pienaar, U. de V. (1966). *The reptiles of the Kruger National Park*. Nat. Parks Brd. of Trustees, Pretoria.

Pitman, C. R. S. (1974). *A guide to the snakes of Uganda. Revised edition*. Wheldon and Wesley pp. 290.

Plessis, S. S. du (1945). Cranial anatomy and ontogeny of the South African cordylid *Chamaesaura anguina. S. Afr. J. Sci. 41*, 245-268.

Poguda, A. A. (1972). The stability of the solutions of the snake toxins from *Vipera lebetina, Echis carinatus* and *Naja oxiana* venom during storage. *Biol. Nauk. 15*, 52-55.

Poilleux, G. and Boquet, P. (1972). Proprietes de trois toxines isolees du venin d'un Elapidae: *Naja melanoleuca. C. r. hebd. Seanc. Acad. Sci. Paris 274*, 1953-1956.

Pooley, A. C. (1962). The Nile crocodile. *Lammergeyer 2*, 1-55.

Pooley, A. C. (1969). Preliminary studies on the breeding of Nile crocodile *Crocodylus niloticus* in Zululand. *Lammergeyer 10*, 22-44.

Porter, K. R. (1972). *Herpetology*. W. B. Saunders, London. pp. 524.

Porter, W. P. and James, F. C. (1979). Behavioral implications of Mechanistic Ecology II. The African Rainbow lizard, *Agama agama. Copeia* 594-619.

Power, J. H. (1930). On the South African species of the genus *Zonurus*. *Ann. Transvaal Mus. 14*, 11-19.

Prato, A. del (1895). Vertebrati Eritri aggiunta al Catalogo della Collezione Eritrea Bottego. *Atti. Soc. Ital. Sci. Nat. Milan 35*, 17-26.

Preiss, F. (1977). *Saurodactylus mauritanicus*, ein Zwerggecko aus Marokko. *Aquar. Terr. Z. (Stuttgart) 30*, 391-392.

Pringle, J. A. (1954). The cranial development of certain South African snakes and the relationships of these groups. *Proc. zool. soc. London 123*, 813-865.

Pringle, J. A. (1955). A new subspecies of the spitting cobra *Naja nigricollis* Reinhardt from the Cape Province. *Ann. Natal Mus. 13*, 253-254.

Pritchard, P. C. H. (1967). *Living turtles of the World.* T. F. H., pp. 288.

Pritchard, P. C. H. (1971). The leatherback or leathery turtle *Dermochelys coriacea. I. U. C. N. Monogr. 1*, 1-39.

Proctor, J. B. (1929). On the remarkable Gecko *Palmatogecko rangei* Andersson. *Proc. zool. soc. London* (for 1928) 917-922.

Rabb, G. B. and Snedigar, R. (1960). Notes on feeding behavior of an African egg-eating snake. *Copeia* 59-60.

Radovanovic, M. (1935). Anatomische Studien am Schlangenkopf. *Jena Z. Natur. 69*, 321-421.

Radovanovic, M. (1967). Phylogenie und Evolution der Giftschlangen. *Zool. Anz. 179*, 199-229.

Raina, R. K., Nganga, J. N., Njoroge, D. K. and Telang, B. V (1977a). Further studies on the mechanism of vadodepressor response in cats after intravenous administration of venom from the snake *Dendroaspis jamesoni* (Jamesons mamba). *Toxicon 15*, 561-570.

Raina, R. K., Nganga, J. N. and Talang, B. V. (1978). Neuromuscular, cardiac and smooth muscle effects of whole venom of *Naja mossambica pallida. Agressologie 19*, 23-28.

Raina, R. K., Njoroge, D. K., Nganga, J. N. and Telang, B. V. (1977b). Centrally induced vasodepressor response after intravenous administration of whole venom of *Naja mossambica pallida* in cats. *Experienta 33*, 1499.

Raj, U. (1976). Incubation and hatching success in artifically incubated eggs of the hawksbill turtle, *Eretmochelys imbricata* (L.). *J. exp. mar. Biol. Ecol. 22*, 91-99.

Rand, A. S. (1958). A new subspecies of *Chamaeleo jacksoni* Boulenger and a key to the species of three-horned chamaeleons. *Mus. Comp. Zool. Breviora 99*, 1-8.

Rand, A. S. (1961). A suggested function of the ornamentation of East African Forest chamaeleons. *Copeia* 411-414.

Rand, A. S. (1963). Notes on the *Chamaeleo bitaeniatus* complex. *Bull. Mus. Comp. Zool. 130*, 1-29.

Raw, L. R. G. (1973). A review of the dusky-bellied water snake, *Lycodonomorphus laevissimus* (Gunther), with descriptions of two new subspecies. *Ann. Natal Mus. 21*, 713-718.

Raw, L. R. G. (1976). A survey of the dwarf chamaeleons of Natal, South Africa, with descriptions of three new species (Sauria: Chameleonidae). *Durban Mus. Novitates 11*, 139-161.

Raw, L. R. G. (1978a). A further new dwarf chameleon from Natal, South Africa (Sauria: Chameleonidae). *Durban Mus. Novitates 11*, 265-269.

Raw, L. R. G. (1978b). Taxonomic notes on the hinged terrapins, genus *Pelusios*, of Natal. (Testudinata: Pelomedusidae). *Durban Mus. Novitates 11*, 287-294.

Reid, H. A. (1977). Prolonged defibrination syndrome after bite by the carpet viper, *Echis carinatus. Brit. med. J. 2*, 1326.

Reitz, C. J. (1977). Berg adder bite. *S. Afr. med. J. 51*, 528.

Rendahl, H. (1939). Zur Herpetologie der Seychellen. *Zool. Jahrb. Syst. 72*, 155-328.

Reymond, A. (1956). Contribution a l'etude de l'action du venin de *Vipera lebetina* (L.) *Trav. Inst. Sci. Cherifien (ser. zool.) 9*, 1-112.

Rhodin, A. G. J., Mittermeier, R. A., Gardner, A. L. and Medem, F. (1978). Karyotypic analysis of the *Podocnemis* turtles. *Copeia* 723-728.

Richards, J. (1979). The flat lizards (*Platysaurus*). *S. W. H. S. Bull. 2*, 11-14.

Richter, H. (1933). Das Zungenbein und seine Muskulatur bei den Lacertilia vera. *Jena Z. Natur. 66*, 395-480.

Rieppel, O. (1979). The braincase of *Typhlops* and *Leptotyphlops* (Reptilia: Serpentes). *Zool. J. Linn. Soc. 65*, 161-176.

Rieppel, O. and Labhardt, L. (1979). Mandibular mechanics in *Varanus niloticus* (Reptilia: Lacertilia). *Herpetologica 35*, 158-163.

Roberts, J. S. and Schmidt-Nielsen, K. (1966). Renal ultrastructure and excretion of salt and water by three terrestrial lizards. *Amer. J. Physiol. 211*, 476-486.

Robertson, L. A. D., Chapman, B. M. and Chapman, R. F. (1965). Notes on the biology of the lizards *Agama cyanogaster* and *Mabuya striata striata* collected in th Rukwa Valley, Southwest Tanganyika. *Proc. zool. soc. London 145*, 305-320.

Robertson, S. S. D. and Delpierre, G. R. (1969). Studies on AFrican snake venoms. IV. Some enzymatic activities in the venom of the boomslang *Dispholidus typus. Toxicon 7*, 189.

Robinson, M. D. (1978). Sexual dichromatism in the Namaqua chamaeleon, *Chamaeleo namaquensis, Madoqua 11*, 81-83.

Robinson, M. D. (1979). Karyology, phylogeny and biogeography of the Namaqua chamaeleon, *Chamaeleo namaquensis* Smith 1831 (Chamaeleonidae), Reptilia). *Beaufortia 28*, 153-156.

Robinson, M. D. and Cunningham, A. B. (1978). Comparative diet of two Namib Desert sand lizards (Lacertidae). *Madoqua 11*, 41-53.

Rochebrune, A. T. de (1884). *Faune de la Senegambie. Reptiles.* Paris, 1-221.

Romer, A. S. (1956). *Osteology of the Reptiles.* Univ. Chicago Press, 1-772.

Rose, W. (1950). *The reptiles and amphibians of Southern Africa.* Cape Town pp. 378.

Rossman, D. A. (1976). Taxonomic status of the West African colubrid snake *Natrix firestonei. Herpetologica 32*, 256-257.

Rossman, D. A. and Eberle, W. G. (1977). Partition of the genus *Natrix*, with preliminary observations on evolutionary trends in natricine snakes. *Herpetologica 33*, 34-43.

Roux, J. (1907a). Sur quelques Reptiles sud-africains. *Revue suisse Zool. 15*, 75-86.

Roux, J. (1907b). Beitrage zur Kenntnis der fauna von Sud-Afrika. Ergebnisse einer Reise von Prof. Max Weber im Jahre 1894. VII. Lacertilia. (Eidechsen). *Zool. Jahrb. Syst. Jena 25*, 403-444.

Roux, J. (1939). Recoltes de R. Paulian et A. Villiers dans le Haut Atlas marocain, 1938 (troiseime note). *Bull. Soc. Sci. nat. Maroc 19*, 11-22.

Roux-Esteve, R. (1969). Les serpents de la region de Lamto (Cote d'Ivoire). *Ann. Univ. Abidjan (ser. E) 2*, 81-140.

Roux-Esteve, R. (1970). Influence de l'atitude sur la variation intraspecifique de *Typhlops angolensis* (BOCAGE) et de *Typhlops lineolatus* JAN (Serpentes, Typhlopidae). *C. R. hebd. seanc. Acad. Sci. Paris D270*, 2185-2186.

Roux-Esteve, R. (1974). Revision systematique des Typhlopidae d'Afrique. Reptilia-Serpentes. *Mem. Mus. nat. Hist. nat. Paris (Nov. ser. A., zool) 87*, 1-313.

Roux-Esteve, R. (1975). Recherches sur la morphologie, la biogeographie et la phylogenie des Typhlopidae d'Afrique. *Bull. I. F. A. N. A36*, 428-508.

Roux-Esteve, R. and Guibe, J. (1965a). Etude comparee de *Boaedon fuliginosus* (Boie) et *Boaedon lineatus* D. et B. (Ophidiens). *Bull. I. F. A. N., A27*, 397-409.

Roux-Esteve, R. and Guibe, J. (1965b). Contribution a l'etude de genre *Boaedon. Bull. Mus. natn. Hist. nat. Paris 36*, 761-774.

Ruppell, E. W. P. E. S. (1835). *Neue Wirbelthiere zu der Fauna von Abyssinien gehorig. III. Amphibien.* Frankfurt pp. 18.

Russell, A. P. (1977). The genera *Rhoptropus* and *Phelsuma* (Reptilia: Gekkonidae) in southern Africa: A case of convergence and a reconsideration of the biogeography of *Phelsuma. Zoologica Africana 12,* 393-408.

Russell, A. P. (1978). The status of the lizard genera *Blaesodactylus* Boettger and *Homopholis* Boulenger (Reptilia: Gekkonidae). *Copeia* 25-29.

Saiff, E. (1975). Preglottal structures in the snake family Colubridae. *Copeia* 589-592.

Saint Girons, H. (1953a). Notes d'ecologie sur les Reptiles du Haut Atlas. *Bull. Soc. zool. France 78,* 13-24.

Saint Girons, H. (1953b). Une vipere naine: *Vipera latastei montana. Bull. Soc. zool. France 78,* 24-28.

Saint Girons, H. (1967). Le cycle sexuel et les correlations hypophyso-genitales des males chez *Agama bibroni* Dumeril au Maroc. *Bull. Biol. France belgique 101,* 321-344.

Saint Girons, H. (1977). Systematique de *Vipera latastei latastei* Bosca, 1878 et description de *Vipera latastei gaditana,* subsp. n. (Reptilia, Viperidae). *Rev. suisse Zool. 84,* 599-607.

Saint Girons, H. (1978). Morphologie externe comparee et systematique des Viperes d'Europe (Reptilia, Viperidae). *Revue suisse Zool. 85,* 565-595.

Saint Girons, H. and Detrait, J. (1978). Communautes antigeniques des venins et systematique des Viperes Europeennes. Etude immunoelectrophoretique. *Bull. soc. Zool. France 103,* 155-166.

Saint Girons, H. and Saint Girons, M. C. (1956). Cycle d'activite et thermoregulation chez les reptiles (lezards et serpents). *Vie Milieu 7,* 133-226.

Savage, J. M. (1952). The correct generic names for the Iguanid lizards of Madagascar and the Fiji Islands. *Copeia* 182.

Schaefer, N. (1970). A new species of house snake from Swaziland, with notes on the status of the two genera *Lamprophis* and *Boaedon. Ann. Cape Prov. Mus. Nat. Hist. 8,* 205-208.

Schaefer, N. (1975). Fang replacement in the snakes *Dispholidus typus* and *Thelotornis kirtlandii. Herpetologica 31,* 102-104.

Schenkel, E. (1901). Achter Nachtrag zum Katalog der herpetologischen Sammlung des Basler Museums. *Verh. naturf. Ges. Basel 13,* 142-199.

Schieck, A., Kornalik, F. and Habermann, E. (1972). The prothrombin-activating principle from *Echis carinatus* venom. I. Preparation and biochemical properties. *Naunyn-Schmiedeberg Arch. exp. Path. Pharmak. 272,* 402.

Schmidt, H. (1966). *Agama atricollis* subsp. inc aus der Serengeti. *Salamandra 2,* 57-68.

Schmidt, K. P. (1919). Contributions to the herpetology of the Belgian Congo based on the collection of the American Museum Congo Expedition, 1909-1915. Part I. Turtles, crocodiles, lizards and chamaeleons. *Bull. Amer. Mus. nat. Hist. 39,* 385-624.

Schmidt, K. P. (1923). Contributions to the herpetology of the Belgian Congo based on the collection of the American Museum Congo Expedition, 1909-1915. Part II. Snakes, with field notes by Herbert Lang and James P. Chapin. *Bull. Amer. Mus. nat. Hist. 49,* 1-146.

Schmidt, K. P. (1933). The reptiles of the Pulitzer-Angola Expedition. *Ann. Carnegie Mus. 22,* 1-15.

Schmidt, K. P. (1943). Amphibians and reptiles from the Sudan. *Field Mus. Nat. Hist. (ser. zool.) 24,* 331-338.

Schmidt, K. P. and Inger, R. F. (1957). *Living reptiles of the World.* Doubleday, N. Y. pp. 287.

Schmidt, K. P. and Marx, H. (1957). Results of the Namru-3 southeastern Egypt Expedition 1954. II. Reptiles and amphibians. *Bull. Zool. Soc. Egypt 13*, 16-28.

Schmidt-Nielsen, K., Borut, A., Lee, P. and Crawford, E. (1963). Nasal salt excretion and the possible function of the cloaca in water conservation. *Science 142*, 1300-1301.

Sclater, W. L. (1898). List of the reptiles and amphibians of South Africa, with descriptions of new species. *Ann. S. Afr. Mus. 1*, 95-111.

Scortecci, G. (1928a). Una nova specie di *Hemidactylus* dell'Eritrea: *Hemidactylus fossatti. Att. Soc. Ital. Sci. Nat. Milano 67*, 33-36.

Scortecci, G. (1928b). Rettili dell'Eritrea esistenti nelle Collezioni del Museo Civico de Milano. *Atti. Soc. Ital. Sci. Nat. Milano 67*, 290-339.

Scortecci, G. (1929a). Una nova specie di *Hemidactylus* dell'Eritrea: *Hemidactylus Zolii. Atti. Soc. Ital. Sci. Nat. Milano 68*, 116-120.

Scortecci, G. (1929c). Primo contributo alla conoscenza dei rettili e degli anfibi della Somalia italiana. *Atti. Soc. Ital. Sci. Nat. Milano 68*, 245-279.

Scortecci, G. (1930a). Rettili e Anfibi raccolti dal Prof. E. Zavattari in Eritrea. *Atti. Soc. Ital. Sci. Nat. Milano 69*, 193-217.

Scortecci, G. (1930c). Contributo alla conoscenza dei rettili e degli anfibi della Somalia, dell'Eritrea e dell'Abissinia. *Boll. Mus. Zool. Univ. Torino (ser. 3) 41*, 1-26.

Scortecci, G. (1930e). Nuove specie di rettili ed anfibi del Mozambico e della Somalia Italiana. *Atti. Soc. Ital. Sci. Nat. 69*, 319-321.

Scortecci, G. (1931b). Secondo contributo alla conoscenza dei Rettili della Somalia italiana. *Atti. Soc. Ital. Sci. Nat. Milano 70*, 127-152.

Scortecci, G. (1932a). Descrizione preliminare di un nuovo ofidio ed un anfibio della Somalia italiana. *Atti. Soc. Ital. Sci. Nat. Milano 71*, 58-60.

Scortecci, G. (1932b). Nuove specie di amfibi e di rettili della Somalia italiana. *Atti. Soc. Ital. Sci. Nat. Milano 71*, 264-269.

Scortecci, G. (1933e). Descrizione preliminare di nuove specie e sottospecie del genere *Pristurus* della Somalia italiana. *Atti. Soc. Ital. Sci. Nat. Milano 72*, 242-244.

Scortecci, G. (1935a). Un nuova genere e una nuova specie di Colubridi Opistoglifi della Penisola del Somali. *Ann. Mus. Stor. nat. Genova 59*, 1-5.

Scortecci, G. (1935c). Il genre *Pristurus* nella Somalia italiana. *Atti. Soc. Ital. Sci. Nat. Milano 74*, 118-156.

Scortecci, G. (1935d). Rettili raccolti nel Deserto Libico dalla Missione Desio della Reale Accademia d'Italia. *Atti. Soc. Ital. Sci. Nat. Milano 74*, 185-190.

Scortecci, G. (1935e). Rettili raccolti nel Deserto Libico dal Prof. Lodovico di Caporiacco. *Atti. Soc. Ital. Sci. Nat. Milano 74*, 191-194.

Scortecci, G. (1939). Spedizione Zoologica del Marchese Saverio Patrizi nel Basso Guiba e nell'Oltregiuba, Rettili, Ofidi. *Ann. Mus. civ. Stor. nat. Giacomo Doria Genova 58*, 263-291.

Seshadri, C. (1956). Urinary excretion in the Indian house lizard, *Hemidactylus flaviviridis* (Ruppell). *J. Zool. Soc. India 8*, 63-78.

Seshadri, C. (1957). Water conservation in *Uromastix hardwickii* (Gray), with a note on the presence of mullerian ducts in the male. *J. Zool. Soc. India 9*, 103-113.

Seshadri, C. (1959). Structural modification of the cloaca of *Lycodon aulicus aulicus*, Linn., in relation to urine excretion and the presence of sexual segment in the kidney of male. *Proc. Nat. Inst. Sci. India B25*, 271-278.

Seurat, L. G. (1934). Etudes zoologiques sur la Sahara central. Mission du Hoggar. III. (Fevrier a Mai 1928). *Mem. Soc. Hist. nat. Afr. Nord 4*, 1-198.

Shayer-Wollberg, M. and Kochva, E. (1967). Embryonic development of the venom apparatus in *Causus rhombeatus*. (Viperidae, Ophidia). *Herpetologica 23*, 249-259.

Shipman, W. H. and Pickwell, G. V. (1973). Venom of the yellow bellied sea snake (*Pelamis platurus*). Some physical and chemical properties. *Toxicon 11*, 375-377.

Shipolini, R., Bailey, G. S. and Banks, B. E. C. (1974). The seperation of a neurotoxin from the venom of *Naja melanoleuca* and the primary sequence determination. *Eur. J. Biochem. 42*, 203-211.

Sill, W. D. (1968). The zoogeography of the Crocodilia. *Copeia* 76-88.

Simbotwe, M. P. (1980). Reproductive biology of the skinks *Mabuya striata* and *Mabuya quinquetaeniata* in Zambia. *Herpetologica 36*, 99-104.

Simroth, H. (1888). Zur Kenntnis der Azorenfauna. *Arch. Naturg. 54*, 179-234.

Sims, K. J. and Singh, I. (1978). Breeding the West African dwarf crocodile, *Osteolaemus tetraspis tetraspis*, at Kuala Lumpur Zoo, with observations on nest construction. *Inter. Zoo. Yrbk. 18*, 83-84.

Sjongren, S. J. (1945). Uber die Embryona lentwicklung des Sauropsidenmagens. *Acta anat. Suppl. 2*, 1-223.

Skinner, J. H. (1959). Ontogeny of the breast-shoulder apparatus of the S. African lacertilian *Microsaura pumila pumila* (Daudin). *Ann. Univ. Stellenbosch A35*, 5-66.

Smit, A. L. (1949). Skedelmorfologie en-kinese van *Typhlops delalandi* Schlegel. *S. Afr. J. Sci. 45*, 117-140.

Smith, A. (1829). Contributions to the natural history of South Africa. *Zool. J. 4*, 433-444.

Smith, A. (1834-1849). *Illustrations of the zoology of South Africa.* London pp. 28 pl. 78. (See Waterhouse, 1880, for exact dates of publication)

Smith, H. M., Smith, R. B. and Sawin, H. L. (1977). A summary of snake classification. (Reptilia, Serpentes). *J. Herpet. 11*, 115-121.

Smith, M. A. (1926). *Monograph of the sea-snakes* (Hydrophiidae). London pp. 130.

Snedigar, R. and Rokosky, E. J. (1949). Notes on new-born gaboon vipers. *Copeia* 145-146.

Sokolov, V. E. (1966). Water content in tissues of some desert animals. *Zool. Zhur. 45*, 776-777.

Somani, P. and Arora, R. B. (1962). Mechanism of increased capillary permeability induced by *Echis carinatus* (saw-scaled viper) venom: A possible new approach to the treatment of viperine snake poisoning. *J. Pharm. Pharmac. 14*, 394-395.

Spence, J. M. (1966). Observations on the Damara Chamaeleon, *Microsaura damarana* Boulenger. *Ann. Cape Prov. Mus. 5*, 145-148.

Spoczynska, J. O. I. (1972). Rearing hatchlings of *Chelonia mydas*. *Brit. J. Herpet. 4*, 263-266.

Staley, F. H. (1929). A case report of Gaboon viper poisoning with recovery. *Bull. Antivenin Inst. Amer. 3*, 31-39.

Stebbins, R. C. (1961). Body temperatures studies in South African lizards. *Koedoe 4*, 54-67.

Steindachner, F. (1907). Uber *Homopholis erlangeri* (n. sp.) aus Abessinien, *Alestes sadleri* Blgr. aus dem Victoria-Nyansa und *Varicorhinus tornieri* (n. sp.) aus Deutsch-Kamerun. *Ann. Naturh. Hofmus Wien 21*, 149-155.

Steindachner, F. (1911). Vorlaufiger Bericht uber drei neue arten aus der familie der Chamaeleontidae. *Anz. Akad. Wiss. Wien 48*, 177-179.

Stejneger, L. (1891). Description of a new scincoid lizard from East Africa. *Proc. U. S. Natl. Mus. 14*, 405-406.

Stejneger, L. (1894). Description of a new species of blind-snake (Typhlopidae) from the Congo Free State. *Proc. U. S. Natl. Mus. 16*, 709-710.

Stejneger, L. (1936). The generic name of a South African ophidian. *Copeia* 114.

Stemmler, O. (1971). Auffallige Schwanzhaltung bei zwei Riesenschlangen (*Python sebae* und *Epicrates cenchria*). *Salamandra 7*, 137-142.

Sternfeld, R. (1908a). Die Schlangenfauna von Kamerun. *Mitt. zool. Mus. Berlin 3*, 397-432.

Sternfeld, R. (1908b). Die Schlangenfauna Togos. *Mitt. zool. Mus. Berlin 4*, 207-236.

Sternfeld, R. (1912a). Reptilia. In, *Wissenschaftliche Ergebnisse der Deutschen Zentral-Afrika Expedition, 1907-1908*. Leipzig *4*, 187-279.

Sternfeld, R. (1912c). Der formenkreis des *Chamaeleon bitaeniatus*. *Sitz. Ges. Naturf. Freunde Berlin 9*, 379-384.

Sternfeld, R. (1917). Reptilia und Amphibia. In, *Ergebnisse der Zweiten Deutschen Zentral-Afrika Expedition, 1910-1911*. Leipzig *1*, 407-509.

Steward, J. W. (1971). *The snakes of Europe*. David and Charles pp. 238.

Stewart, M. M. (1968). Notes on reptiles from Northern Malawi. *J. Herpet. 2*, 163-164.

Steyn, W. (1962). Verdere lig- en elektronemikroskopie van die derde oog met aanmerkings oor sy termoregulatiewe funksie. *Cimbebasia 4*, 7-16.

Steyn, W. (1963). *Angolsaurus skoogi* (Andersson). A new record from South West Africa. *Cimbebasia 6*, 8-11.

Steyn, W. and Els, A. J. (1963). *Python anchietae* Bocage: a note on prey capture and diet. *Cimbebasia 6*, 16-22.

Steyn, W., Finkeldey, H. and Buys, P. J. (1963). *Agama hispida makarikarica* Fitzsimons. A preliminary note on its occurrence in South West Africa and behaviour. *Cimbebasia 6*, 12-15.

Steyn, W. and Haacke, W. D. (1966). A new webfooted gecko (*Kaokogecko vanzyli* gen. et sp. nov.) from northwestern South West Africa. *Cimbebasia 18*, 1-23.

Steyn, W. and Mitchell, A. J. L. (1965). A new Scincid genus, and a new record from South West Africa. *Cimbebasia 12*, 2-12.

Steyn, W. and Mitchell, A. J. L. (1967). Two new gekkos from South West Africa. Cimbebasia 21, 9-21.

Stimson, A. F. (1969). Liste der rezenten Amphibien und Reptilien. Boidae (Boinae + Bolyeriinae + Loxoceminae + Pythoninae). *Das Tierreich 89*, 1-49.

Strydom, A. J. C. and Botes, D. P. (1971). Snake venom toxins. Purification, properties and complete amino acid sequence of two toxins from ringhals (*Hemachatus haemachatus*) venom. *J. biol. Chem. 246*, 1341-1349.

Strydom, D. J. (1972). Snake venom toxins. The amino acid sequence of two toxins from *Dendroaspis polylepis* (black mamba) venom. *J. biol. Chem. 247*, 4029-4047.

Strydom, D. J. (1977a). Snake venom toxins. The amino acid sequence of toxin Vi2, a homologue of pancreatic trypsin inhibitor, from *Dendroaspis polylepis polylepis* (black mamba) venom. *Biochim. biophys. Acta 491*, 361-369.

Strydom, D. J. (1977b). Snake venom toxins. The amino acid sequence of a short-neurotoxin homologue from *Dendroaspis polylepis polylepis* (black mamba) venom. *Eur. J. Biochem. 76*, 99-106.

Stuart, C. T. (1976). A note on *Pythonodipsas carinata* Gunther (Western keeled snake) in South West Africa, with new distribution records. *Madoqua 9*, 63-64.

Sukhanov, V. B. (1961). Some problems of the phylogeny and systematics of Lacertilia. *Zool. Zhur. 40*, 73-83. (in Russian).

Sulter, M. M. (1962). A contribution to the cranial morphology of *Causus rhombeatus* (Lichtenstein) with special reference to cranial kinesis. *Ann. Univ. Stellenbosch A37*, 1-40.

Sweeney, R. C. H. (1960). The Chelonia of Nyasaland Protectorate. *Nyassal. J. 13*, 35-50.

Sweeney, R. C. H. (1971). *Snakes of Nyasaland, with new added corrigenda and addenda*. Asher, Amsterdam pp. 200.

Switak, K. H. (1966). Notes on the nutrition and care of the Madagascan day gecko *Phelsuma madagascariensis*, at Steinhart Aquarium. *Inter. Zoo Yrbk. 6*, 107.

Szczerbak, N. N. (1971). Taxonomy of the genus *Eremias* (Sauria, Reptilia) in connection with the focuses of the desert-steppe fauna development. *Paleoarctic Vest. Zool. 2*, 48-55.

Szczerbak, N. N. (1974). *The Palearctic Desert Lizards.* Acad. Sci. Ukranian S. S. R. Inst. Zool., Kiev pp. 296 (in Russian).

Szczerbak, N. N. and Golubev, M. L. (1977). Systematics of the Palearctic geckos (genera *Gymnodactylus, Bunopus, Alsophylax*) *Proc. Zool. Inst. Acad. Sci. U. S. S. R. 74*, 120-133.

Taborska, E. (1971). Intraspecies variability of the venom of *Echis carinatus. Physiol. bohemoslov 20*, 307-318.

Taborska, E., Mebs, D. and Kornalik, F. (1978). Pharmacological and biological properties of venom gland extract from an opistoglyphous snake *Thelotornis kirtlandii. Toxicon 16*, 428 (Abst.)

Tansley, K. (1959). The retina of two nocturnal geckos, *Hemidactylus turcicus* and *Tarentola mauritanica. Pflugers Arch. ges. Physiol. 268*, 213-220.

Tansley, K. (1961). The retina of a diurnal gecko *Phelsuma madagascariensis longinsulae. Pflugers Arch. ges. Physiol. 272*, 262-269.

Teichner, O. (1978). Breeding the West African dwarf crocodile, *Osteolaemus tetraspis tetraspis* at Metro Toronto Zoo. *Inter. Zoo Yrbk. 18*, 88-89.

Tercafs, R. R. (1962). Observations ecologiques dans le massif du Tibesti (Tchad). *Rev. Zool. Bot. Afric. 66*, 107-126.

Theakston, R. D. G., Lloyd-Jones, M. J. and Reid, H. A. (1977). Micro-elisa for detecting and assaying snake venom and venom-antibody. *Lancet* 639-641.

Thominot, A. (1878). Note sur un nouveau genre de Reptile de la Famille des Geckotiens. *Bull. Soc. Philom. Paris (ser. 7) 2*, 254-256.

Thominot, A. (1887). Description de trois especes nouvelles d'*Anolis* et d'un amphisbaenien. *Bull. Soc. Philom Paris (ser. 7) 11*, 182-190.

Thorpe, R. S. (1975a). Biometric analysis of incipient speciation in the ringed snake *Natrix natrix* (L.). *Experientia 31*, 180-182.

Thorpe, R. S. (1975b). Quantitative handling of characters useful in snake systematics with particular reference to intraspecific variation in the ringed snake *Natrix natrix* (L.) *Biol. J. Linn. Soc. 7*, 24-43.

Thorpe, R. S. (1979). Multivariate analysis of the population systematics of the ringed snake, *Natrix natrix* (L.) *Proc. Roy. Soc. Edinburgh B78*, 1-62.

Thorpe, R. S. (1980). A comparative study of ordination techniques in numerical taxonomy in relation to racial variation in the ringed snake *Natrix natrix* (L.) *Biol. J. Linn. Soc. 13*, 7-40.

Thorpe, R. S. and Crawford, C. M. (1979). The comparative abundance and resource partitioning of two green-gecko species (*Phelsuma*) on Praslin, Seychelles. *Brit. J. Herpet. 6*, 19-24.

Thorpe, R. S. and McCarthy, C. J. (1978). A preliminary study, using multivariate analysis of a species complex of African house snakes (*Boaedon fuliginosus*). *J. Zool. London 184*, 489-506.

Throckmorton, G. S. (1976). Oral food processing in two herbivorous lizards, *Iguana iguana* (Iguanidae) and *Uromastyx aegyptus* (Agamidae). *J. Morph. 148*, 363-390.

Tilak, R. (1964). The osteology of *Uromastyx hardwickii* Gray. *Zool. Anz. 173*, 403-426.

Tornier, G. (1900a). Neue liste der Crocodile, Schildkroten und Eidechsen Deutsch Ost-Afrikas. *Zool. Jahrb. Syst. 13*, 579-618.

Tornier, G. (1901a). Die Crocodile, Schildkroten und Eidechsen in Togo. *Arch. Naturg. Festschrift 67,* 65-88.

Tornier, G. (1902a). Herpetologisch neues aus Ost-Afrikas. *Zool. Anz. 25,* 700-703.

Tornier, G. (1902b). Herpetologische neues aus Deutsch Ost-Afrika. *Zool. Jahrb. Syst. 15,* 578-590.

Tornier, G. (1902c). Die Crocodile, Schildkroten und Eidechsen in Kamerun. *Zool. Jahrb. Syst. 15,* 663-677.

Tornier, G. (1905). Schildkroten und Eidechsen aus Nordost-Afrika und Arabien. *Zool. Jahrb. Syst. 22,* 365-388.

Toye, S. A. (1972). The locomotory activity of the Rainbow lizard *Agama agama* (L.). *Brit. J. Herpet. 4,* 252-257.

Traill, T. S. (1843). *Essay on the physiognomy of Serpents (Schlegel, H.). Translated by Thos. Stewart Traill.* Edinburgh pp. 25.

Trost, E. (1953). Die Entwicklung, Histogenese und Histologie der Epiphyse, der Paraphyse, der Velum transversum, des Dorsalsackes und des subcommissuralen Organs bei *Anguis fragilis, Chalcides ocellatus* und *Natrix natrix. Acta anat. 18,* 326-342.

Tu, A. T., Lin, T. S. and Bieber, A. L. (1975). Purification and chemical characterization of the major neurotoxin from the venom of *Pelamis platurus. Biochemistry, N. Y. 14,* 3408-3413.

Underwood, G. (1954). On the classification and evolution of geckos. *Proc. zool. soc. London 124,* 469-472.

Underwood, G. (1967). A contribution to the classification of snakes. *Brit. Mus. Nat. Hist. 653,* 1-179.

Underwood, G. (1968). On the status of some South African vipers. *Ann. Cape Prov. Mus. (Nat. Hist.) 6,* 81-85.

Underwood, G. (1971). A modern appreciation of Camp's "Calssification of the Lizards". In, *Camps Classification of the lizards.* S. S. A. R. Facsimile reprint, vii-xvii.

Underwood, G. (1976). A systematic analysis of boid snakes. In A. d'A. Bellairs and Cox, C. B., *Morphology and biology of reptiles.* Academic Press, 151-175.

Urban, E. K. (1970). Nesting of the green turtle (*Chelonia mydas*) in the Dahlak Archipelago, Ethiopia. *Copeia* 393-394.

Valenciennes, A. (1861). Note sur les Animaux d'Abyssinie rapportes par M. Courbon. *Comptes Rendus Acad. Sci. Paris 52,* 433-434.

Valverde, J. A. (1957). *Aves del Sahara espanol. Estudio ecologico del desierto.* Madrid pp. 487.

Van Dam, G. P. F. (1921a). Descriptions of new species of *Zonurus* and notes on the species of *Zonurus* occuring in the Transvaal. *Ann. Transvaal Mus. 7,* 239-243.

Van Dam, G. P. F. (1921b). Description of a new variety of a South African lizard of the family Gekkonidae. *Ann. Transvaal Mus. 7,* 244.

Van Den Heever, J. A. (1976). The cranial and cervical muscles of the South African limbless lizard *Typhlosaurus aurantiacus aurantiacus* Peters (Reptilia, Sauria). *Ann. S. Afr. Mus. 69,* 169-214.

Van Der Walt, S. J. (1972). Studies on *Bitis arietans* venom. IV. Association of protease A. *Z. phys. Chem. 353,* 1217-1227.

Van Der Walt, S. J. and Joubert, F. J. (1971). Studies on puff adder (*Bitis arietans*) venom. I. Purification and properties of protease A. *Toxicon 9,* 153-161.

Van Der Walt, S. J. and Joubert, F. J. (1972). Studies on the venom of puff adder (*Bitis arietans*). II. Specificity of protease A. *Toxicon 10,* 341-356.

Vandeventer, T. L. and Schmidt, M. (1977). Caesarean section on a western gaboon viper, *Bitis gabonica rhinoceros. Inter. Zoo Yrbk. 17,* 140-142.

Vanlennep, E. W. and Komnick, H. (1970). Fine structure of the nasal salt gland in the desert lizard *Uromastyx acanthinurus*. *Cytobiologie 2*, 47-67.

Vanzolini, P. E. (1951). A systematic arrangement of the family Amphisbaenidae (Sauria). *Herpetologica 7*, 113-123.

Varaldi, M. (1951). Sauriens recueillis dans la partie Sud-Ouest du Maroc. *C. R. Soc. Sci. nat. Maroc 17*, 105.

Vasilescu, E. (1970). Isolated head of the tortoise *Emys orbicularis*. *Rev. Roum. biol. zool. 15*, 273-276.

Versluys, J. (1898). Die mittlere und aussere Ohrsphare der lacertilia und Rhynchocephalia. *Zool. Jb. Abt. Anat. 12*, 161-406.

Vesey-Fitzgerald, D. F. (1948). Reptiles and Amphibians from the Seychelles Archipelago. *Ann. Mag. Nat. Hist. (ser. 11) 14*, 577-584.

Vesey-Fitzgerald, L. D. E. F. (1958). The snakes of Northern Rhodesia and the Tanganyika borderlands. *Proc. Trans. Rhod. scient. Ass. 46*, 17-102.

Viljoen, C. C. and Botes, D. P. (1979). Influence of pH on the kinetic and spectral properties of phospholipase A_2 from *Bitis gabonica* (Gaboon adder) snake venom. *Toxicon 17*, 77-87.

Viljoen, C. C., Botes, D. P. and Schabort, J. C. (1975). Spectral properties of *Bitis gabonica* venom phospholipase A_2 in the presence of divalent metal ion, substrate and hydrolysis products. *Toxicon 13*, 343.

Viljoen, C. C., Meehan, M. and Botes, D. P. (1979). Seperation of *Bitis gabonica* (Gaboon adder) venom arginine esterases into kinin-releasing, clotting and fibrinolytic factors. *Toxicon 17*, 145-154.

Viljoen, C. C., Schabort, J. C. and Botes, D. P. (1974). *Bitis gabonica* venom. A kinetic analysis of the hydrolysis by phospholipase A_2 of 1,2-dipalmitoyl-sn-glycero-3-phosphorylcholine. *Biochim. biophys. Acta 360*, 156.

Viljoen, C. C., Visser, L. and Botes, D. P. (1976). An essential tryptophan in the active site of phospholipase A_2 from the venom of *Bitis gabonica*. *Biochim. biophys. Acta 438*, 424.

Viljoen, C. C., Visser, L. and Botes, D. P. (1977). Histidine and lysine residues and the activity of phospholipase A_2 from the venom of *Bitis gabonica*. *Biochim. biophys. Acta 483*, 107.

Villiers, A. (1950a). Contribution a l'etude du peuplement de la Mauritanie. Ophidiens. *Bull. I. F. A. N. 12*, 984-998.

Villiers, A. (1950b). Contribution a l'etude de l'Air. Reptiles ophidiens et cheloniens. *Mem. I. F. A. N. 10*, 337-344.

Villiers, A. (1958). Tortues et crocodiles de l'Afrique Noire Francaise. *Init. afr. I. F. A. N. 15*, 1-354.

Villiers, A. (1963). Les serpents de l'ouest Africain. *Init. afr. I. F. A. N. 2* (ed. 2), 1-190.

Vinciguerra, D. (1928). Risultati zoologica della Missione inviata della R. Societa Geografica Italiana per l'esplorazione dell'oasi di Giarabub (1926-1927). Rettili, Batraci e Pesci. *Ann. Mus. civ. Stor. nat. G. Doria 52*, 324-345.

Vinciguerra, D. (1931). Spedizione scientifica all'oasi di Cufra. Rettili. *Ann. Mus. civ. Stor. nat. G. Doria 55*, 248-258.

Vinson, J. and Vinson, J. M. (1969). The saurian fauna of the Mascarene Islands. *Mauritius Inst. Bull. 6*, 203-320.

Visser, J. (1967a). Color varieties, brood size and food of South Africa *Pelamis platurus* (Ophidia: Hydrophiidae). *Copeia* 219.

Visser, J. (1967b). First report of ovoviviparity in a Southern African amphisbaenid *Monopeltis c. capensis*. *Zool. Africana 3*, 111-113.

Visser, J. and Chapman, D. S. (1978). *Snakes and snakebite. Venomous snakes and management of snakebite in Southern Africa*. Purnell pp. 152.

Visser, J. G. J. (1972). Ontogeny of the chrondrocranium of the chamaeleon *Microsaura pumila pumila* (Daudin). *Ann. Univ. Stellenbosch A47,* 1-68.

Vogel, Z. (1964). *Reptiles and amphibians. Their care and behaviour.* Studio Vista, London pp. 228.

Voris, H. K. (1977). A phylogeny of the sea snakes (Hydrophiidae). *Fieldiana Zoology 70,* 79-166.

Wahlstrom, A. (1971). Purification and characterization of phospholipase A from the venom of *Naja nigricollis. Toxicon 9,* 45-56.

Warburg, M. R. (1964). Observations on the microclimate in habitats of some desert vipers in the Negev, Arava and Dead Sea regions. *Vie Milieu 15,* 1017-1041.

Warrell, D. A. and Arnett, C. (1976). The importance of bites by the saw-scaled or carpet viper (*Echis carinatus*): epidemiological studies in Nigeria and a review of the world literature. *Acta Trop. Sep. 33,* 307-341.

Warrell, D. A., Barnes, H. J., and Piburn, M. F. (1976). Neurotoxic effects of bites by the Egyptian cobra (*Naja haje*) in Nigeria. *Trans. Roy. Soc. Trop. Med. Hyg. 70,* 78-79.

Warrell, D. A., Mc. Davidson, N., Greenwood, B. M., Ormerod, L. D., Pope H. M., Watkins, B. J. and Prentice, C. R. M. (1977). Poisoning by the bites of the saw-scaled or carpet viper (*Echis carinatus*) in Nigeria. *Quart. J. Med. (new ser.) 46,* 33-62.

Warrell, D. A., Mc. Davidson, N., Ormerod, L. D., Pope, H. M., Watkins, B. J., Greenwood, B. M. and Reid, H. A. (1974). Bites by the saw-scaled or carpet viper (*Echis carinatus*): trial of two specific antivenoms. *Brit. Med. J. 4,* 437-440.

Warrell, D. A., Greenwood, B. M., Mc. Davidson, N., Ormerod, L. D. and Prentice, C. R. M. (1976). Necrosis, haemorrhage and complement depletion following bites by the spitting cobra (*Naja nigricollis*). *Quart. J. Med. 45,* 1-22.

Warrell, D. A. and Ormerod, L. D. (1976). Snake venom ophthalmia and blindness caused by the spitting cobra (*Naja nigricollis*) in Nigeria. *Amer. J. Trop. Med. Hyg. 25,* 525-529.

Warrell, D. A., Ormerod, L. D. and Mc. Davidson, N. (1975). Bites by puff adder (*Bitis arietans*) in Nigeria, and value of antivenom. *Brit. J. Med.* 697-700.

Warrell, D. A., Ormerod, L. D. and Mc. Davidson, N. (1976). Bites by the night adder (*Causus maculatus*) and burrowing vipers (genus *Atractaspis*) in Nigeria. *Amer. J. Trop. Med. Hyg. 25,* 517-524.

Waterhouse, F. H. (1880). On the date of publication of the parts of Sir Andrew Smith's "Illustrations of the zoology of South Africa." *Proc. zool. soc. London* 489-491.

Webb, M. (1951). The cranial anatomy of the South African geckos *Palmatogecko rangei* (Andersson), and *Oedura karroica* (Hewitt). *Ann. Univ. Stellenbosch A27,* 131-165.

Welch, K. R. G. (1980). A comment on the European colubrid genus *Haemorrhois. S. W. H. S. Bull. 3,* 14-15.

Wermuth, H. (1953). Systematik der rezenten Krokodile. *Mitt. zool. Mus. Berlin 29,* 375-514.

Wermuth, H. (1965). Liste der rezenten Amphibien und Reptilien. Gekkonidae, Pygopodidae, Xantusiidae. *Das Tierreich 80,* 1-246.

Wermuth, H. (1967). Liste der rezenten Amphibien und Reptilien. Agamidae. *Das Tierreich 86,* 1-127.

Wermuth, H. (1968). Liste der rezenten Amphibien und Reptilien. Cordylidae (Cordylinae + Gerrhosaurinae). *Das Tierreich 87,* 1-30.

Wermuth, H. (1969). Liste der rezenten Amphibien und Reptilien. Anguidae, Anniellidae, Xenosauridae. *Das Tierreich 90,* 1-42.

Wermuth, H. (1977). Liste der rezenten Amphibien und Reptilien. Testudines, Crocodylia, Rhynchocephalia. *Das Tierreich 100,* 1-174.

Wermuth, H. and Fuchs, K. (1978). *Bestimmen von Krokodilen und ihrer Haute.* Verlag, Stuttgart pp. 100.

Werner, F. (1897). Ueber Reptilien und Batrachier aus Togoland, Kamerun und Tunis aus dem Kgl. Museum fur Naturkunde in Berlin. *Verh. zool.-bot. Ges. Wien 47,* 395-408.

Werner, F. (1898). Uber reptilien and batrachier aus Togoland, Kamerun und Tunis II. *Verh. zool.-bot. Ges. Wien 48,* 191-231.

Werner, F. (1899). Ueber reptilien und batrachier aus Togoland, Kamerun und Deutsch-Neu-Guinea, grosstentheils aus dem Kgl. Museum fur Naturkunde in Berlin. *Verh. zool.-bot. Ges. Wien 49,* 132-157.

Werner, F. (1907). Ergebnisse der mit Subvention aus der Erbschaft Treitl unternommenen zoologischen Forschungsreise Dr. Franz Werner's nach dem aegyptischen Sudan und Nord-Uganda. XII. Reptilien und Amphibien. *Sitz. K. Akad. Wiss. Math.-Nat. Kl. Wien 116,* 1823-1926.

Werner, F. (1915). Reptilia und Amphibia. In, Michaelson, W., *Beitrage zur Kenntnis der Landund Susswasser-fauna Deutsch Sudwestafrikas. 3,* 325-376.

Werner, Y. L. (1964). Frequences of regenerated tails, and structure of caudal vertebrae in Israeli desert geckos (Reptilia: Gekkonidae). *Israel J. Zool. 13,* 134-136.

Werner, Y. L. (1977). Ecological comments on some gekkonid lizards of the Namib Desert, South West Africa. *Madoqua 10,* 157-169.

Wever, E. G. (1968). The ear of the chamaeleon: *Chamaeleo senegalensis* and *Chamaeleo quilensis. J. Exp. Zool. 168,* 423-436.

Whaler, B. C. (1971). Venom yields from captive gaboon vipers (*Bitis gabonica*). *Uganda J. 35,* 195-206.

Whaler, B. C. (1972). Gaboon viper venom and its effects. *J. Physiol. London 222,* 61.

Whaler, B. C. (1975). Cardiovascular and respiratory effects of gaboon viper venom. *Gen. Pharmac. 6,* 35-41.

Wiedersheim, R. (1876). Zur Anatomie und Physiologie des *Phyllodactylus europaeus* mit besonderer Berucksichtigung des Aquaeductus vestibuli. *Morph. Jahrb 1,* 495-534.

Wiegmann, A. F. A. (1834). *Herpetologia Mexicana.* Berolini pp. 54.

Wiegmann, A. F. A. (1837). Herpetologische Notizen. *Arch. Naturg. 3,* 123-136.

Wiley, E. O. (1979). Ventral gill arch muscles and the interrelationships of gnathostomes with a new classification of the Vertebrata. *Zool. J. Linn. Soc., 67,* 149-179.

Willemse, G. T. and Hattingh, J. (1979). Physiological effects of fresh freeze-dried and commercially prepared rinkals (*Hemachatus haemachatus*) venom. *Toxicon 17,* 89-93.

Willemse, G. T., Hattingh, J. and Coetzee, N. (1979). Precipitation of human blood clotting factors by puff-adder (*Bitis arietans*) venom. *Toxicon 17,* 331-335.

Willemse, G. T., Hattingh, J., Karlsson, R. M., Levy, S. and Parker, C. (1979). Changes in composition and protein concentration of puff-adder (*Bitis arietans*) venom due to frequent milking. *Toxicon 17,* 37-42.

Wilson, D. (1959). The hatching and rearing of the cobra *Naja melanoleuca. Brit. J. Herpet. 2,* 159-162.

Wilson, V. J. (1965). The snakes of the Eastern Province of Zambia. *Puku 3,* 149-170.

Witte, G. F. de (1922a). Description de reptiles nouveaux de Congo Belge. *Rev. Zool. Africaine 10,* 66-71.

Witte, G. F. de (1922b). Description d'un Ophidien nouveau recolte au Congo par le Dr. Schouteden. *Rev. Zool. Africaine 10*, 318-319.

Witte, G. F. de (1930a). Un serpent nouveau du Congo Belge (*Rhinocalamus rodhaini* sp. n.) *Rev. Zool. Africaine 19*, 1-3.

Witte, G. F. de (1930c). Missions Saharienne Angerias-Draper 1927-1929. Reptiles et Batraciens. *Bull. Mus. Nat. Hist. nat. paris 2*, 614-618.

Witte, G. F. de (1933a). Description de deux amphisbaenides nouveaux du Congo Belge. *Rev. Zool. Africaine 23*, 168-171.

Witte, G. F. de (1933b). Description de Reptiles nouveaux provenant du Katanga (1930-31). *Rev. Zool. Africaine 23*, 185-191.

Witte, G. F. de (1933c). Description d'un *Typhlops* nouveau provenant du Ruanda (*Typhlops lestradei* SP. N.). *Rev. Zool. Africaine 23*, 206.

Witte, G. F. de (1941). Exploration du Parc National Albert, Mission G. F. de Witte (1933-1935). Batraciens et Reptiles. *Inst. Parcs Nat. Congo Belge 33*, 1-261.

Witte, G. F. de (1953). Exploration du Parc National de l'Upemba, Mission G. F. de Witte. Reptiles. *Inst. Parc Nat. Congo Belge 6*, 1-322.

Witte, G. F. de (1954). Revision des Amphisbaenidae du Congo Belge. *Vol. Jubilaire V. von Straelen, Brussels 2*, 981-1010.

Witte, G. F. de (1959). Contribution a la faune herpetologique du Congo Belge. Description de trois serpents nouveaux. *Rev. Zool. Africaine 60*, 348-351.

Witte, G. F. de (1962). Genera des serpents du Congo et du Ruanda-Urundi. *Mus. Roy. Afr. Ter., Sci. zool. (ser. 8) 104*, 1-203.

Witte, G. F. de (1963). The colubrid snake genera *Chamaelycus* Boulenger and *Oophilositum* Parker. *Copeia* 634-636.

Witte, G. F. de (1964). A new chamaeleon from the Congo. *Amer. Mus. Novitate 2192*, 1-3.

Witte, G. F. de and Laurent, R. F. (1942a). Contribution a la systematique des Amphisbaenidae du Congo Belge. *Rev. Zool. Africaine 36*, 67-86.

Witte, G. F. de and Laurent, R. F. (1942b). Contribution a la faune herpetologique du Congo Belge. *Rev. Zool. Africaine 36*, 101-115.

Witte, G. F. de and Laurent, R. F. (1943a). Contribution a la systematique des Boiginae du Congo Belge (Reptilia). *Rev. Zool. Africaine 37*, 157-189.

Witte, G. F. de and Laurent, R. F. (1943b). Contribution a la systematique des formes degradees de la famille des Scincidae, apparentees au genre *Scelotes* Fitzinger. *Mem. Mus. Roy. Hist. nat. belge (ser. 2) 26*, 1-44.

Witte, G. F. de and Laurent, R. F. (1947). Revision d'une groupe de Colubridae Africains. *Mem. Mus. Roy. Hist. nat. belge (ser. 2) 29*, 1-134.

Wood, S. C. and Johansen, K. (1974). Respiratory adaptions to diving in the Nile monitor lizard *Varanus niloticus. J. Comp. Physiol. 89*, 145-158.

Wood, S. F. (1938). Variations in the cytology of the blood of geckos, *Tarentola mauritanica*, infected with *Haemogregarina platydactyli*, *Trypanosoma platydactyli* and *Pirhemoctyon tarentolae. Univ. California Publ. Zool. 41*, 9-21.

Woodward, S. F. (1933). A few notes on the persistence of active spermatazoa in the African night adder, *Causus rhombeatus. Proc. zool. soc. London* 189-190.

Zavattari, E. (1909). Muscoli ioidei dei sauri in rapporto con i muscoli ioidei degli altri vertebrati. *Mem. Acad. Sci. Turin 61*, 55-135 and 351-392.

Zeiller, W. (1969). Maintenance of the yellow bellied sea snake *Pelamis platurus* in captivity. *Copeia* 407-408.

Zimmerman, H., Habermann, E. and Lasch, H. G. (1971). Der Einflu von Gift der Sandrasselotter (*Echis carinatus*) auf die Hamostase. *Thromb. Diath. Haemorrh. 25*, 425-437.

INDEX TO GENERA

INDEX TO SPECIES AND SUBSPECIES